The Call of the Sagas

THIS BOOK IS PART OF ISLINGTON READS BOOKSWAP SCHEME

Please take this book and either return it to a Bookswap site or replace with one of your own books that you would like to share.

If you enjoy this book, why not join your local Islington Library and borrow more like it for free?

Find out about our FREE e-book, e-audio, newspaper and magazine apps, activities for pre-school children and other services we have to offer at www.islington.gov.uk/libraries

Pekka Piri

The Call of the Sagas
Finland to Iceland in an Open Boat

Translated from the Finnish by Owen F. Witesman

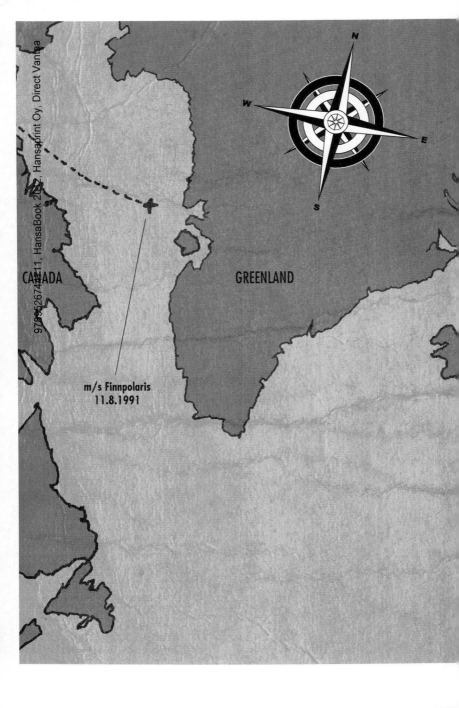

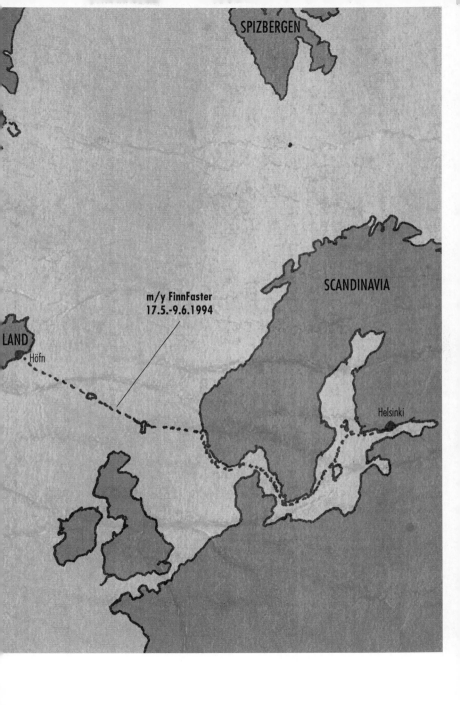

First English edition

Photographers
Matti Murto (first picture)
Anton Soinne (pictures 2-4)
Sveinn Björnsson (last picture)
Pekka Piri and Matti Pulli

© WSOY 1996 and 1997, PIRIUS 2010 (Finnish), 2012 (English)

Cover by Jarkko Nikkanen, Lapine Oy
Cover picture by Anssi Asunta
Layout and typography by Jarkko Nikkanen, Lapine Oy

ISBN 978-952-67447-1-1

FILI The translation of this work was supported by a grant from FILI—Finnish Literature Exchange

Owen F. Witesman asserts his moral right to be identified as the translator of the work.

Preface to the English Edition

The translation of *The Call of the Sagas* has been a continuation of the journey described within these pages. Overcoming the challenges along the way would not have been possible without the help and support of many dedicated friends and professionals. For their financial support of this project I wish especially to thank FILI—Finnish Literature Exchange and its director, Iris Schwanck. I would also like to thank the translator, Owen Witesman, both for his hard work and friendship. Thanks also to Soila Lehtonen, Kate Lambert, and Hanna Wagner for their editorial comments. Veijo Meri, Mirkka Rekola and Kalle Holmberg provided needed inspiration, as did Markku Niskala and Rita Landström. The genuine encouragement of my sons, Perttu and Arttu, for their father's strange endeavor has been invaluable.

I dedicate this translation to the memory of my dear friend and navigator, Matti Pulli, who sailed to his eternal rest on June 18, 2011.

Pekka Piri, Helsinki, June 9, 2012

To the Reader

A tiny splinter, a tantalizing idea took up residence in my mind. At times it faded and receded, only to surge back and begin to grow anew. It took on shape and form, beginning to dominate my thoughts. It grew into a dream. Finland to Iceland in an open boat!

Mad, stupid bravado! Unbelievable lack of judgment, megalomania, sensationalism. I heard these criticisms, but I didn't care. Not everyone understands everything. They probably said the same to Leif Ericson and Erik the Red, just like they did to Columbus . . . But still they went.

Amidst the circus at our port of departure, a reporter asked how you get the call of the sagas, where something like that comes from. My short answer was that it comes from an extremely turbulent life. This answer still holds true, one year and a half later. A turbulent life comes from a hunger for experiences where you set your own limits.

A journey is a crossing of borders, in the final analysis the blasting away of the borders within ourselves. The length of the journey is determined by the distance to your own horizon.

As I was starting the powerful outboard motor of the seven-meter open boat m/y *FinnFaster* with my friend, sea captain Matti Pulli, next to me as navigator, no one could tell us our dream would come true. We couldn't even tell ourselves. But we had our own firm faith in our experience, endurance, and equipment. These aren't things you ask a consultant about—you make the evaluations yourself when it's your own life on the line.

Matti and I will share this experience, just the two of us, for our entire lives. Our progress across the Baltic, the North Sea, and the North Atlantic, 2,179 nautical miles from the Helsinki Market Square to Höfn in Iceland is a long voyage. Long out in nature and on the map, but even longer in the human mind, experienced personally.

Our difference in temperament and accepting that difference bore us to our destination. It became an indispensable driving force in our small boat as it struggled across the ocean. We were finally rewarded when after 23 days on the 9th of June at 2:42 PM, Europe's largest glacier, Vatnajökull, rose above our horizon.

I would like to thank everyone who supported this project, who helped in outfitting and launching the *FinnFaster*. I feel genuine triumph in the face of all the unbelievers who closed doors, bullied, and mocked.

Without Juha Snell's forthright trust in us, my plan, and the seaworthiness of the *FinnFaster* boat he had built, the Call of the Sagas probably never would have begun.

My close, true friends encouraged the realization of the dream. I didn't converse about seafaring—navigation, steering techniques, or fuel economy—with them. We talked about much more important things: crises, dreams, daring—life. Thanks to Jussi Iivonen, Markku Niskala,

Lassi Mäkinen, and many, many others for all of your support.

My thoughts turn warmly to a pair of unassuming, salt water-bitten professional seamen, Navigation Officer Hemming Karlström and Shipmaster Lasse Eriksson on barren Kökar. On many a winter night I sat with them in the light of an oil lamp pondering my plan amidst the symphony of a winter storm. They believed in it firmly from the very beginning. They never asked, "Why on earth would you do that?" Some people understand that sometimes you just have to go—sometimes far away.

Turning that starter key was also made decidedly easier by the fact that our families did not question our motives or our capacity as we cast off from the Estonia Basin stone pier in Helsinki's South Harbor, heading 2,200 nautical miles into the unknown.

Tapaninvainio, Helsinki, December 13, 1995
Pekka Piri

zzz

I have arrived. I have made my journey again in my mind, in memories and thoughts. The justification for the existence of *The Call of the Sagas* as a book is that it is true. I have described one episode in the life of humanity as accurately as I am able. I hope that the book will give the reader more than the experience of a boat trip.

For sharing the joy and pain of writing, I would like to thank Jussi Iivonen most particularly. Without his encouragement and support to be open, and without him patiently listening to the text, the result might have been only a story.

I dedicate my book to my sons, Perttu and Arttu. They have experienced the joy and pain of *The Call of the Sagas* close up. Their adulthood is before them, and they are my sons.

April 14, 1996

TABLE OF CONTENTS

Mayday, Mayday, Mayday 17
The Outermost Islet 19
I Want a Divorce 26
The Map 30
The Herring Market 44
So Let's Build a Boat 52
Absolutely 55
Yes or No 61
The Blue Letter 64
Goodbye, Mom 70
Can You Dig It? 72
Muru 74
How about Going to Harmaja First? 78
After the Night, Morning 83
Hell 89
No More Kidding Around 96
I Just Want to Help 104
The Joy of the Surf 111
Vivi 121
I Couldn't Trust Anyone 126
Damn It, It Won't Work 131
Moon Shadow 140
Shut Your Trap Already 148
Thanks, Hans! 152
You're Never Coming Back 155
The Pilot of Chicago 158
Rán's Daughters 161
Pantyhose 169
Hold On 173
Do It Again 178
The Scent of Summer and Winter 183

My Bergen 186
A Pitstop and Edvard Grieg 189
The Special Nightgown 198
Leaving Scandinavia 200
Keep Counting for a Minute More 207
I'm Sorry 211
The Blue Hour 219
Storm Island Maija 225
The Column 227
The Fog is my Friend 233
He Rules the Air and Elements 235
Haven't We Already Said Everything? 238
Mykines 244
The Interrogation 251
Location Infinity 255
Howdy 264
Fulmarus glacialis 268
Is It True? 270
The Sigurvindur 277
It Won't Stop! 280
I Love that Woman 289
Matkro 291
Full Faith and Credit 297
The Leader 301
Do You See any Motorboat Down There? 304
Welcome Back 308
Old Mustachioed Boys 315
The Iron Ring 320

POSTSCRIPT 322
PICTURES 324
YEARS LATER 331
APPENDIXES 341

Far away

Clouds over the sea,
waves: History
*   A distant kingdom—*
*   I will go beyond the seas*

A thousand-year wake
following Leif and Erik
*   pulls me, and I feel:*
*   I too will find something*

Amidst the fog an island—
lava, fire, and ice
*   Over the roar a shout:*
*   Friends, understand*

I am leaving. I am leaving,
far beyond the ancient waves
*   Why, why? Where?*
*   To follow the call!*

Before start, in March 1994

Mayday, Mayday, Mayday

The captain of the Finnish freighter *Finnpolaris* makes his decision. The distress call, "Mayday, mayday, mayday, *Finnpolaris*," hurtles into the ether on the morning of August 11, 1991. In rough seas an ice floe weighing hundreds of tons has ripped gaping holes in her right side. The nearest land is a hundred nautical miles to the east, in Greenland.

The gray, four-meter swells lick ever higher at the starboard side. The 18-man crew and one passenger have donned their survival suits—the vessel has begun to list severely to the right.

The captain orders the lifeboat down.

"It just had to happen, goddamn it," he curses in his mind. Did this really have to happen? He already knows that the ship is lost; only saving the crew has any meaning right now. Still, infuriating thoughts about what is to come flash through his mind as it works feverishly: the maritime declaration, the interviews, and all the paperwork, paper on and on . . . the experts and judges, standing wisely on the shore. . . . Bugger it, it's time to work. Every last one of us has to survive. This is going to be an endurance test. No one's nerve can fail. This is still my job. . . .

The first mate has lowered the boat and jumps in after it into the surging ocean. The captain rushes to the bridge once more to notify the shore radio that the ship has been abandoned and then returns to the edge of the sinking right side with his fists clenched. The 160-meter, only ten-year-old ocean freighter only has a little time left.

"Are we sure everyone is in the boat?"

"Yes, yes. Get off there already!" The lifeboat rolls violently out into the frigid grayness of the empty ocean. The captain does not think of his ship anymore; her death is inevitable. The mass of steel shaped by skilled engineers and containing the latest technology is slipping into the deep. The bow is almost entirely under the surface, the right side is covered by water, and the waves are already washing across the deck. The vessel is sinking bow first, the name no longer visible. The pressure rips the hatches off and the zinc concentrate cargo explodes a hundred meters into the gray Arctic sky.

The last vortex left by the sinking ship has disappeared. In the empty haze nineteen people look on silently at the empty spot in the sea where the ship that had been carrying them only a few hours earlier disappeared. The *Finnpolaris* had gone to the bottom.

The Outermost Islet

The August sun beats down on Kökar, a small island located in the middle of the northern Baltic Sea. I'm satisfied about the results of my morning's work: many more meters of weather boarding are up. My sweat is the good sweat of work.

I'm hard at work on a new scheme: a small fishing cabin, built warm enough for the winter, at the base of a rocky bay. To the south, a fantastic view of the Baltic Sea opens up between the sheltering islands, unmarred all the way to Gotland; to the north a crag rises dozens of meters above sea level. Seal hunters lived here three thousand years ago in their huts of skins, one thousand years ago the Vikings set out from here for the East, six hundred years ago the Franciscan monks came with their hymns and monasteries.

People have departed here and people have arrived here, with the only road being the Baltic. Seafarers on their successful distant voyages, unlucky castaways, fugitives, unnamed men, smugglers with their precious cargoes, trying to lose customs officers . . . lives, fates in a lonely, miniature community that is nevertheless connected to anywhere in the world where the sea extends.

This has been my place for more than a decade. I came here with my family for the first time on our own adventure-seeking expedition with a car, a trailer, a small boat, and a tent. Our two boys with their hedgehog haircuts seeking Viking treasure, Elina to enjoy the abundant flora and sun, me trying to get as far away as possible, physically and spiritually, from the absurd world of my work. The fatigued personnel manager fished for food for his family, swam with his browned boys on a distant islet in the open sea, made friends with the local fishermen, learned to know their life, and was accepted from the very first moment. We returned summer after summer. I fled here in the autumn and even in the snow and ice, despite only having a tent to weather the winter storms. What I found here was an authentic life and people waiting for me to return whom I could respect and who as people had become very close to me.

The joy of progress at the building site fills my mind, even though my life is in anything but good order. My profession of nearly two decades has fallen by the wayside, at once a relief and an aching bruise.

"Burn-out," they say, so civilized; the most civilized know how to say it in English too. It's all the same what the diagnosis is. Of course I know that these things happen—generally. But not to me in particular? Hell no! Bad things—tragedies—aren't supposed to happen to you personally. You read about them in the newspaper, all nice and tidy, like death. Of course they exist somewhere "generally," happening to other people, but not right here, to me, to my family. But it's still true, unassailably.

I'm building my cabin so I can have something new, something spiritually my own, and I couldn't make my hideout in any more natural a place. I'm building on my good friend Harry's lot, on posts poured thirty years ago by his father, Nestor, the old fisherman, at an ancient Viking refuge.

One more meter of board and I dash off to the store. The familiar boat, the powerful motor, the no-nonsense, long-distance equipment, everything in balance, fitted with my own hands and well maintained. Water flies from the sides, the sun beats down, and the building site recedes.

What are they up to at the firm right now? I imagine myself in the familiar runaround, clarifying the supplemental clauses of the signature protocol of a complicated collective labor agreement written up by the lawyers as the CEO rages on in his own personal nightmare drama and the chief shop steward hints threateningly about the upcoming personnel budget discussions. Hundreds of people forced to put on a show for each other, deceiving each other and themselves and guarding their interests because hundreds of mortgages are crying out to be paid. The working world has become a ghoulish court with all of its attendant slippery lackeys. I see the bright professionals I hire quickly growing depressed and feel guilty for luring them in. My throat constricts and my mouth goes dry; these feelings are too familiar—this lack of any hope has gone on too long.

I decide to take the afternoon off; this heat won't last forever. Instead of heading toward the building site, I point the bow towards Kökarören, the outermost islet about ten miles to the south. At the beginning of the century, during the fishing high season, as many as four hundred people lived cheek by jowl in their shacks on this one hundred by two hundred meter island each autumn during the fishing season. Now there is only one shack left, thankfully restored.

I sit with my lunch on the south edge of Ören. The normal south-westerly sea wind pushes the swells up onto the smooth rock. The smoked flounder and cold beer taste good as I sit there alone with my thousand thoughts.

Why am I here? How have I found my paradise? I know scads of unbelievable places on the sea, and I've usually found them myself. The reason has to be hidden in the fact that I am a greedy wanderer.

Suddenly I comprehend a logic in my wandering: I always have to go to the outermost islet, like now; otherwise I'm only halfway. In my first small boat I had to reach Gråskärsbådan on the approach to Helsinki, then Tammisaari's Segelskär, Ören here in Kökar, and a year ago my blood drew me from the Arctic Ocean to the Atlantic, around Nordkapp. I had to go to the edge, beyond the most remote corner of Europe, to experience the gray immensity and struggle myself in my own simple open boat on the swells of the ocean.

On the outermost islet I have arrived; from there I can see far enough, and it is natural to stop there. And without exception there is space there. You don't have to look at stupid yuppies bumbling around with their fathers' or companies' outfits. On the outermost islet I have only met real people, the kind whom it is a joy to know, who know how to talk about real things, whom you can stand to listen to, and who are humble enough to listen themselves. Out there the whiskey gets passed around, if there happens to be any.

I wander around my islet, imaging the compact life of the fishing community here on this rocky outcrop lashed by the autumn winds. They were not on a jaunt—they earned their bread for their children, for their elderly, and for themselves with the herring they caught. On my first trip to Kökar, an old man told me how it felt as a small boy to see right there, right in front of me, a ship overturn and a whole family drown at once.

The haven next to me has always been used by the seafarer. The fisherman, the seal hunter, the smuggler, the Viking . . . and now me.

My thoughts stop at the Vikings. They went a long way. The cove of this islet was a waypoint for them. I close my eyes and see the cleaving of the waves, hear the clash and the clamor, the rattling of the chains, the shouts and curses of the strong men. In the background rings the laughter of an alluring woman. . . . I smell the scent of sea, tar, blood, and sweat. They appear over there, out of the fog from behind Karlskär and continue far into the east.

I return to the building site after the sun has set. The moon hangs low behind me, the swell churning powerfully, and I enjoy the ease of steering and navigating in the dark velvet. Tomorrow I'll finish the north wall, and then I just have the east side before moving to the interior.

A candle burns in my temporary dwelling, in the old tar-smelling shed on the shore. Swallows nest under its eaves and large vipers dive into the warrens of its stone foundation which extends down into the water. It's good to lounge on Nestor's big fish chest; the hot rum toddy hits the spot, and I soon fall asleep.

Through my languor I hear the news from the crackling radio. ". . . the Finnish cargo ship *Finnpolaris* ran into trouble and sank last night off the western coast of Greenland. At the time of the accident there were a large number of ice floes, rough seas, and fog in the area. The *Finnpolaris*' first distress call was picked up in Halifax, Canada, from which the authorities in Greenland were alerted to initiate rescue operations. The Danish tanker *Sofia* Theresa picked up the crew of the sunken ship after seven hours of drifting on the ocean. . . ."

I think about the crew of the *Finnpolaris*. That was a long seven hours. Now we know they were saved, but when they were in that boat they couldn't know that. My thoughts fly to the next day's newspapers: the harsh headlines, the speculation, the erroneous information, because they have to make a story of it no matter what. I think of the captain of the ship; he has hard times ahead of him. The sea protest, the interviews, the whole machinery being put in motion. I respect the unknown shipmaster for having given the distress call in time; everyone was saved. Will the unlucky captain's friends still be friends after the accident? Based on my experience, I see many who will quickly turn their backs. I feel sympathy for him.

I Want a Divorce

It's bright in our Helsinki row house, the evening light in the yard giving a broad landscape view into the spacious living room. A few logs have been burning in the fireplace as usual; the whole family likes a real fire.

"Where do the linens go?" Elina asks as I hold a sketched diagram of my cabin in my hand.

"Nowhere," I answer. "There aren't any linens at the cabin. You sleep in sleeping bags."

"How can you even imagine something like that. It's dirty."

"Listen, Ellu, I've spent plenty of time living in fishing cabins. I know their world. That's just how it is. I'm not going to build a single linen closet in my cabin. Sleeping bags are enough. If that doesn't agree with someone's idea of hygiene, then they can bring their own bed linens along. This is very simple. I'm not building a summer cottage, and I'm not building a bungalow. Torrskata is going to have a cabin, a proper fishing cabin."

"I can't understand how you can think so simply."

"Come off it, Ellu. No one can tell me how I should want my fishing cabin to be. Linens? What a laugh. I've heard this tune enough times in my life. If you want clean sheets, you carry them in your own two hands. Is that clear, damn it? I'm not just pissing around with this project!"

Elina is quiet. I can see I've beaten her down again, but I'm not going to compromise my idea. I'm going to build a proper fishing cabin, not some trumped up excuse for a summer house. I won't give an inch; I can't violate my building site—it's too unique. Nestor didn't even have a proper sleeping bag.

We aren't capable of a constructive discussion, always two opposing forces squaring off with our antlers. Our life together has been going poorly for years anyway; we squabble frequently. We both suffer from it, and my job loss has only made the difficulties worse. Our shared silences are growing longer and there is a constant strain in our conversations. We take everything we can get whenever we can. It's oppressive. The future is completely open and there is no intimacy. We're trapped in a painful emptiness.

The words catch in my throat; I can't complete the thought. I hear the bitter laughter, try to say something, to explain, but then the anger overcomes me and I explode again. I lash out at Elina, my mind full of anguish at being prevented from drawing near, at being prevented from seeking mutual understanding.

I'm standing next to the fireplace. My soul rages again, and the wine bottle explodes on the back wall of the fireplace.

Why the hell can't I even try, why do we always have to look for explanations for all this mishmash in the easiest possible way! Images crash through my mind: the sleepless nights, the agonizing arguments, the sulking, the silence, the wasted life.

The cinders hiss, the wine boils on the embers, doors slam, and we have to be able to get up in the morning to get the boys to school. Damn, damn, damn, damn!

I heard and saw her say it perfectly clearly: "I want a divorce."

The strained face in the door to the kitchen said it; I heard it, but didn't think about it. I was only thinking about the pain of this messed-up life. The declaration meant the same thing as any of the other messages that had become so familiar: I don't like you, I don't want to be near you. But I'm still responsible for things; that's how it's been for the past thirty years—I'm worried about how she'll get by . . . no, it must have just been another way for her to lash out at me.

"Pekka, we're done. I want a divorce. I can't stand being with you. This is over. It has to be."

She's serious. I see it and I understand—I know. I've known for a long time now. I'm beaten. I hear everything, believing and not believing. The end? Our home, our life together, our whole familiar routine, our whole foundation? I don't know how to think that way. I'm the father of this family after all, on the front lines whenever I'm needed. You can't revoke that. This isn't made to be torn down, this home of ours, this refuge, this nest. It can't be true. The charges in our cannons are just too large. Our home can't be destroyed from the inside out!

Silence, distance, and cynicism have entered to stay. I often find myself wanting to be somewhere else, somewhere very far away with enough distance from absolutely everything, from this whole complicated, unjust, petty, petit bourgeois fairytale of a system. I long for something real, something made to my own measurements; I've become restless and irritable, not feeling like I can fit in the narrow space allotted to me, that I can walk the prim paths others have laid out for me.

The Map

I unroll the package, excited even though I know what it contains. I've finally bought it—I'm done flipping through school maps. I spread the North Atlantic sea chart out on my desk and look at it, alone. I hope that no one will surprise me. I want to be alone with my secret. My thoughts are far away.

There they are, solemn in their solitude in the bosom of the ocean: Shetland, the Faroe Islands, and, farthest off, Iceland. Islands lined up in the North Atlantic, the route along which Leif Ericson and Erik the Red sailed a thousand years ago.

Again I hear the roar, the rattling, the shouts, and the creaking of the ropes. Again the bows froth and somewhere the hips of an ample woman sway. The grayness of the ocean and the eternal swell surround the lonely sea travelers; the spirit of the sagas hangs in the fog.

Pelle Krause, the original good old boy from Maritim, sold me that map like any other one before. He couldn't know what was pounding inside of me. I wanted to really see my dream, in the language of seafarers; I wanted to see exactly, to count the miles, to think and to work. Am I serious? I don't know how to answer that, but I had to buy the map. My stubborn thought won't let up; it just keeps making my blood run quicker.

To Iceland in an open boat! Massive mountain waves before me, sheer immensity, solitude, eternity, fog, and haze.

Exhaustion, struggle, emptiness. Finally the coast of that distant kingdom, the goal, the island of the sagas, lava, fire, and ice. Onshore the stalwart harbor of a fishing village, perhaps even a smoky pub with its broad-fisted, weathered fishermen.

Would it be possible? I measure the route from the right edge, from ancient Bergen of whale hunter lore to the left edge, to Iceland, land of eternal ice. Leg by leg, it is possible! It is possible for me. With the right strategy, equipment, crew, and grit I can do it, but it's a long, fierce journey, really long. And that is exactly why it fascinates, exactly why it makes me so excited, makes my palms sweat . . .

It Just Won't Seem to Rain

A smiling waiter in a white shirt arrives at the table with his notepad.

"Ah, it looks like you've already decided on your order?"

"Yes indeed. We'll have four rib eye steaks, medium, a bottle of Mouton Cadet and two mineral waters to drink, and why not Camparis for everyone."

"Thank you."

The early summer view from the Captain's Table on Sirpalesaari Island is captivating. The light of the setting sun paints the lighthouse tower on Harmaja, and to the south-west rest the familiar silhouettes of Pihlajasaari and Melkki Islands. A few sailboats are gliding into the harbor, pushed by the rising sea wind.

I think the location is appropriate; I was the one who suggested it, and it also suited Elina. The mood is perhaps a little nervous, but still cozy. I look at my sons and feel stirrings of joy and gratitude. Two upstanding young men with open gazes and healthy ambitions, brimming with exuberance. Elina is dressed prettily in blue and has made herself up with understated elegance, as always.

Our small family is celebrating our 25th wedding anniversary. We wanted to mark it somehow, despite everything. You don't brush things like this aside, even though we have grown apart as spouses. We also wanted this moment for our sons, since their existence is based on our union and they are what matters most to both of us.

Even though I feel a little silly, I won't give up; this is our celebration and extremely real. So I stand up and try to speak, to talk to those closest to me.

"Dear Elina . . . boys, Perttu and Arttu . . . I . . ." I'm trying to talk about my feelings—about our struggle, the time we shared, the importance of our children . . . my mouth goes dry, my voice goes hoarse. I am moved by the warmth we still have and nostalgic for what has disappeared.

"This is for you, Elina. Our journey has been a long one; thank you for sticking it out."

I give her a small, silver anchor, a symbol of hope. I don't imagine she'll every wear it, but I couldn't come up with anything else. The anchor—hope—is near and dear to me, but I never give gold. I want to be a realist. Gold isn't my metal—it symbolizes perfection, and I prefer to wear overalls.

We reminisce together about everything fun—the car trips, the boating outings, the ski tours carrying our lunches. . . . We've been busy together. I silently consider the past years; they've been a breath-taking tumble, a fierce, high-stakes run through the rapids. We've gotten through it somehow, but we're exhausted. Once again I remember Muru and swallow, longing. "Everything that doesn't kill us makes us stronger." Muru used to say that a lot.

During our celebration I look out to sea often. We've boated by here dozens of times, the boys small and eager—sometimes in my lap behind the helm, sometimes in the bottom of the simple craft on the way home in a deep, healthy sleep, carefully covered with blankets by Elina. I feel that a page is turning for Elina and me—that it has already turned. All I see ahead is emptiness.

The October weather over the Gulf of Finland is clear. The afternoon sun is setting behind Hanko Peninsula, and I am sitting alone in my boat along the flank of Segelskär, near Finland's southernmost islet. The day's catch is nice, about twenty long-tailed ducks and a few fat common eiders.

The bouillon comes to a boil; I gulp down a couple of cups to warm myself up before the cold journey to Lappvik Harbor.

The bird hunting was a side issue again; I look at the sea again, inquiringly, using my twenty years of experience to consider the possibilities for the Iceland trip. It excites me even more intensely than before, and my thoughts return to it often. I analyze, weigh, calculate, and evaluate the main details. No one has told them to me; I'm inferring everything myself, based on what I've experienced. The thought is as much my own as anything can be—in all its ferocity. The reality of the old Viking route and the distant island of the sagas have begun to take on a life of their own in my mind; adrenaline courses through my veins whenever I imagine arriving at that destination, the distant silhouette outlined above the horizon after an eternity. Oh how that would feel!

The engine is already running; I weigh the anchor, leave the canopy down, and am ready to go. My thoughts are in the fishing village of Höfn on the south-east shore of Iceland. Imagine the mood there as the descendants of the Vikings return from the sea. . . .

My Mercedes lunges along the snowy November road toward Helsinki. I have just seen Muru for the first time after a long silence; I just had to see her. She spoke to me distantly, formally, like a girls' school headmistress lecturing her charges.

I wasn't close to her anymore after all. We did not rejoice together. We did not play and kid each other warmly, trade riddles, or laugh even once. I only heard the official-sounding lecture and felt extraneous. I don't want to believe it. Our phone calls have become formal. She hasn't answered a phone message in a long time.

Muru, who has become closer to me than anyone, who has given me such deep intimacy and real joy despite the distance, has gone somewhere she doesn't want me. I stop at a bus stop and look with empty eyes at the dark forest. For a moment I don't know where I am or why. I feel exhausted. My grief is true.

"Drive to Kaitaistentie. I'll follow behind you. Watch to make sure you don't lose me," I am saying to a driver in the center of Turku late on a winter night. I don't want to be late for our meeting, and I'm unsure of the route. I'm taking a taxi as a pilot.

The name on the door of the red row house reads Karakorpi. A beefy man with an extremely thick beard invites me inside. He seems friendly.

"Heikki Karakorpi. Come on in."

"Pekka Piri. Thank you, I'm glad I could come."

"Oh, it's nothing. Let's get down to business. We can talk it over generally first and then we can look at my pictures. I've tried to put them in order a bit."

By chance I'd heard about a yachtsman in Turku who had been to Iceland the year before, made a note of his name and now I'm sitting in his living room discussing his trip. It's the most valuable information I can get from anyone, despite the fact that we skipper very different boats. Heikki has an open water sailboat, and I pilot a powerful outboard motor boat.

Our phone conversation was interesting. I introduced myself, told him my business, and suggested that we meet. He spoke in a rather roundabout way—not trying to brush me off, but hesitantly.

"The essentials are weather forecasts, navigation in extreme conditions, steering control, and fuel economy. And in terms of the crew, wicked persistence and mental endurance. You can't have agoraphobia showing up in the middle of the North Atlantic."

I know my business. I'm able to explain it concisely because I've gone through it myself, building the concept, and it's my own life that's on the line. That gives the analysis seriousness, a foundation.

Heikki listens intently, nodding now and then. I hear long experience in his comments. He shows dozens of slides and promises to make me copies of every heavy sea picture with precise weather condition information. Then I can study them myself in peace and evaluate the capacity of my equipment.

We part as friends.

"I wouldn't have guessed that you'd have studied this so thoroughly," he says.

"Can't afford not to, especially as an open boat driver," I reply truthfully.

The cabin's oil lamp illuminates the paneled room with soft warmth on a November night. The fireplace is crackling snugly and has once again quickly increased the temperature to a comfortable level. The south-westerly wind is howling at nearly twenty meters per second straight at the porch, but not even a candle flickers. There is no draft here. I'm very satisfied with my hideout; I made sure that it was built well, because I want it to offer shelter long after I am gone. Over the door I slapped a rusted, old Wickström cylinder head, on the wall a couple of seal skins, and on the bunks as an invitation, woolly sheepskins.

I look at Anneli's tender face on the other side of the table. She is always smiling, even when she's serious. I've been marveling at this for years. She has her very own charm.

Anneli knows what I'm considering. She has deigned to listen to my thoughts and understands that out of a tiny fragment of an idea has grown a big dream, and that I'm wrestling with whether it really is possible. She also knows that I've decided to make my decision soon; it can't be put off for long for scheduling reasons.

"Just listen honestly to yourself. You have experience and to spare."

"Yeah, I guess. A lot of times I just think about it from my family's perspective. They have bigger pressures anyway, and are probably afraid of me deciding to go."

"Well, that's clear, but they also know how important it is to you."

We chat for a long time about our dreams, our children, our divorces, and our crises, the whole range of middle-aged life, about the hill we've already climbed. You can see a long ways back from up there and can predict what's ahead as well.

"Listen, Pekka, I'm going to say one thing."
"Yeah?"
"For the same reason I like you so much, I couldn't imagine living with you."

"How so, why not?"

"Because there's so much man in you!"

"Oh. What does 'man' mean to you then?"

"Well, first that you'll just suddenly snatch that horrible officer's knife out of your breast pocket and use it to whack off a monstrous chunk of *gravlax* and then pick it up with your fingers and eat it all at once. Then you throw back a drinking glass full of whiskey in one gulp just to get drunk faster. And then you rush to the shelf, dig out your trumpet, and start playing a solo on the porch so loud it wakes up people all the way in Gotland. That kind of man. And so damned fun!"

The oil lamp is burning with a low flame; I've added wood to the stove. The candles are out, and outside it's pitch-black, the storm still at it.

"Are you sleeping already, Pekka?"

"Mmm . . . almost . . . What was that?"

"I just wanted to say that I hope with all my heart that you and Muru can still find each other again someday."

She nuzzles up close and strokes my beard.

I start getting interested in the sagas and visit the library. I'm curious about everything connected to Iceland, its ancient history and present day. I'm collecting pieces of an unknown puzzle; I want to know more. Behind my curiosity and interest hums the thrilling, true history that has gotten such a strong hold on me. The Vikings have come back to life in my mind.

"... the sagas, fairy tales—what would humans be without fairy tales? It is a great privilege to have them as a legacy ... to us Icelanders the sagas are a priceless treasure, without which we might not even have an independent cultural identity as a community ... when we arrive at certain places we cannot avoid repeating the Icelandic sagas in our minds ... the past lives in the land and in us and the sagas become reality."

I set the book down and think for a long time about Vigdís Finnbogadóttir's profound words, which have broadened my perspective.

The Saga of the Sworn Brothers is especially captivating:

"... Thorgeirr and Thormodr have made an oath by walking under three peat sods that the one who lived the longest would avenge the other. They sail on the Jökulsfjord and meet a cold wind, freezing temperatures, and a snowstorm. The ship turns sideways and water rolls in. Everyone gets wet and their clothing freezes ... Rán's daughters tried the men, inviting them to their embrace. ..."

I figure out that Rán was the wife of the sea god Ægir and they had nine daughters who were the waves of the sea. Rán ruled the kingdom of the dead and all those who drowned became her guests.

I think of the characters of Ægir and Rán, thinking of the twists and turns of human history, which gave rise to those verses, and feel that I would like to meet Rán's daughters myself.

The January day has been clear; the sun even made an appearance. The only thing on my mind is my decision, whether I'm ready. The evening sky is darkening quickly, and I'm sitting at my map again, contemplating. I measure the familiar distances, which I already know from memory, calculating fuel consumption limits, average speeds, and swell patterns. I know the area's weather statistics and ponder what they mean, constructing different danger situations in my mind and then picking them apart.

What am I really thinking about? What's stopping me?

In the blink of an eye I realize that the only obstacle is me. Either I go or I chicken out. Have I done my homework only to chicken out? Is that what I've harnessed all my experience for to do this analysis? No, no, and no! I'm not going to chicken out. I want to experience seeing that shore on my horizon. I'm going!

The boat division of the large conglomerate reacts passively to my memorandum. They don't return my calls. Their negative stance comes out gradually, evasively. I want to know the reasons and hear different, completely contradictory, explanations. I find myself confused and enraged. No one has any obligation to join my undertaking, but honest, straight talk is the least I expect.

I call the company's spokesman, thinking that as a professional he will be able to understand the value of the project. I explain my proposal, again, and receive an answer from very high up: "How is our business of yours?" I'm like someone hit me over the head with a club. Despite my forty-seven years of life and two decades of business experience, it appears I still didn't know that there are some PR representatives you aren't supposed to call!

In February a pilgrimage occurs. The commercial symbol of spring performs a week of rituals in the Helsinki Convention Centre; the boat people gather.

Bright artificial lights in the ceiling illuminate dreams: shining, sleek, colorful dreams which move either by sail of by motor. Smooth, well-spoken presenters wax eloquent about the finer points of these dreams; there is nothing shoddy in this world—at least everything is shiny. Plush seats, teak, warm showers . . . absolutely necessary for the more fortunate in our welfare state.

Many have a fixed, clear dream from year to year: five more feet of length and a little more luxury than before. I don't understand it. They probably look for relaxation and variety from boating, but they want it to be as easy as possible, packaged up like their own living rooms. That's the place for flamboyant sailing caps, blazers, brand-name shoes, and saloons filled with brass toys. Well, there is still plenty of space at sea, just so long as they don't endanger anyone else.

I wander around, lost in my thoughts. I never get very excited about this. My boomerang galls me, especially that when it keeps coming back it isn't accompanied by straight talk. I know what it's all about. It isn't sailing or boating in the background, not marketing or business. We're firmly in the realm of personal chemistry in this jealous land of petty prigs.

I come up next to the conglomerate's tidy stand. I look over the boat I'm ready to set out in. There it sits nicely up on blocks, shined up invitingly. In my mind I start the powerful engine, set my distant course, and begin to listen to the accelerating slapping on the hull. . . .

A young sales manager is leaning on the side of the boat, looking satisfied. I go over to chat.

"There it is. Nice ride. It's too bad my proposal didn't catch any wind. We could have really put it to the test."

The guy perks up; I don't understand what he's doing.

Wordlessly he stretches up his neck, turns it back and forth for a long time as if peering somewhere far off on the horizon, and then says slowly, drawing out each word:

"It . . . just . . . won't seem to rain . . ."

I understand the jibe.

"Nope, doesn't seem that way."

I pat the stern of the boat with my face blank, but my cheek muscles are tense. My brain is boiling; this was all that was missing! Goddamn dickweed! You don't know who you're kidding with. I'm going to shove those words back down your goddamn throat! You're going to swallow them letter by letter and period by period. My plan goes to the grave the day the last rock in this country is pulled from the ground! I've done my homework, what about you? "It just won't seem to rain" . . . to a grown man, damn it!

The Herring Market

In October, life at the Market Square in Helsinki is intense. The herring market is in full swing again. The boats bobbing cheek by jowl and the smell of the fish, the crisp October air, the good humor of the weather-beaten islanders, and the customers stretching their necks in front of the stalls intermingle as a powerful meeting of cultures. A good-humored leisureliness hangs in the damp, limpid air, and the Japanese tourists click their cameras.

I'm in the thick of it, on the deck of the Rex from Kökar helping my friend Harry selling. As a long-term loyal customer and friend of the fishermen, I enjoy the event. I'm with my own people again. Harry often stops by my cabin, now finished, sometimes in his boat, sometimes on his tractor—sometimes with a full load of firewood in the trailer.

Business is good. Expert buyers, tourists, and colorful foreigners bustle about on the dock, humor blossoming in many languages. The traffic on Eteläranta is humming as usual, but distantly. I glance across the street. The familiar main office of my former employer, the familiar front door; I've walked through it many times. All the pin-striped men who hold the world together hurry by, briefcases in hand to their offices in the Palace. They all look the same: an overcoat, a well-tailored suit, and a leather box with a handle.

I know what happens inside: trained lawyers are coaching a large group of company representatives about the wolf traps lurking in the interpretation of collective labor agreements, the management of an enormous system of regulations—catches and tricks—preparing them for the front lines in the impending industrial action and predicting the political policy statements of the other side's union meeting. Lobbying for their interest groups, as they call it, just like on the other side, in Hakaniemi. In those markets they trade in peace in the labor market. This self-perpetuating system oppresses me, a civilized war with memoranda, statistics, and pressure as the weapons, the offices the battleground, day and night, the troops the masses and the unions, the commanders the personnel managers and shop stewards, the ammunition threats, strikes, and lockouts, the victories and defeats percentages, pennies, contract clauses and their interpretations. I glance at the door again. Who could help them? Do they live by what they feel? Does work have to be like that? How many officials and regulators really know they earn their pay? How much cynicism should and can a person put up with?

"Good morning, do you have real salted fish with the heads on?"

An elderly woman wakes me from my reverie. She buys a medium-sized wooden pail full and a loaf of authentic black bread. She knows precisely what she wants and pays exactly from her scant pension money. She talks about the war, about real Baltic herring casserole, about her dear departed Jaska and her children. I'm hearing about a life that has been lived. She is one of the quiet women of our country, whose tenacity has always really been the thing that kept the day-to-day life of the nation going. I slide a jar of spiced fish into her worn, black bag for good measure.

I have hung some of Lasse's sea drawings to the bow of the boat. Lasse is from Kökar too. Secretly I watch the people who look at the pictures, and now and then someone wants to buy one too. Now an elderly man is inspecting them. His reactions are interesting. He has remarkably erect posture, is about average size, and his bearded face is weathered. He is dressed in green outdoors clothing and a red beret.

"The pictures are of Kökar, by Lasse Eriksson, who lives in the area. He knows his subject. Would you like to buy one?"

"We'll see. They do look good. What kind of a man is this Lasse?"

"He's a sailor, a little over fifty years old, a pilot cutter driver, a fisherman, a painter, a ship builder, and a musician, just off the top of my head. A fine fellow. I know him well."

"He certainly knows the sea. You can't help but see that." His comment sticks in my head. What do green-suited, red-beret-wearing men know about the sea? Why was his assessment of Lasse's pictures so spontaneous? The man is somehow sympathetic and seems pretty damn sharp. An interesting fellow. I know tons of people. In my old job I hired hundreds of them. I've used the aptitude tests, and I have a lot of experience in these things, but now I find myself in a surprising situation. This hairy-faced guy doesn't fit into any of my boxes. What kind of a man is he? He isn't just a city slicker playing around in that get-up. The gear is his own, it really belongs to him. The voice and speech tell of natural self-confidence; he's clearly used to responsibility. He isn't an army man; he doesn't fit that mold. Somehow he has a distant look, even though he's strictly present. And besides, he recognized the subtle drawings from a distance and pushed his way closer just because of them. What kind of a man is he and why am I thinking about it?

We're already well into the morning, so I decide to take a break. The red beret has bought his picture. His satchel is on the dock and he's standing looking grimly preoccupied, his arms crossed. He is silent, as if his thoughts were somewhere far away.

"Are you in a rush?"

"Not particularly. How so?"

"Jump in the boat. I'll offer you a beer on the aft deck if you're game."

"Well, why not? Thanks, that would be fine." He strides onto the deck. I can see immediately that he moves with experience, not timid at all. I grab a few beers from the steering cabin and we go aft to sit. It feels nice to rest my feet and get away from the counter for a minute. The seagulls are circling over the Cholera Basin. Harry is throwing more things to sell up out of the hold, and the red beret flips his bottle open with the back of a knife.

We chat about the boat, the market, and the fishermen. He is interested and follows the conversation. He asks if I'm always here and I answer that I am when I can, because it's so nice to be here, that I've never done a better job. We chat more—I can open up easily if I want to, if the thing across from me is a person and not a calculator.

I tell about my experiences in the archipelago, on the salmon lines, about my salmon netting trips with Lasse, about herring trolling in the Gulf of Finland on winter nights, and about the fogs of Nordkapp on the Arctic Ocean. He stays silent, nodding a little. I can feel his sharp gaze, but it doesn't bother me. No one is able to threaten me, and his sharpness is pure attention.

I ask straight up what he does with himself.

"I'm a skipper, have been for a couple of decades."

"In the air or at sea?"

"At sea, as a ship's captain."

"Ah, so you're a professional. In that much time you must have seen a bit of this and that."

"No kidding. I've even had a ship sink out from under me."

I find myself off balance. I know that captains who have experienced shipwrecks don't exactly chat about it; usually they're left with a sort of trauma and would rather keep quiet about it. The red beret told me about it like it was nothing more than a rain shower—me, a complete stranger. I take a chance and continue.

"When did it happen?"

"A little more than two years ago on the Arctic Ocean. We took a gash in the side and had to leave the boat. It was an open-and-shut case. Nothing to do about it. We got the crew into the lifeboat in good order, and luckily everyone survived."

"Wait a second. What was the name of the ship?"

"It was the *Finnpolaris*, a Juliana class bucket, 160 meters. We were taking zinc concentrate from Canada to Louisiana."

"Damn. Listen, I remember that. We heard a lot about it."

"Yeah, it was quite the thing, and you can bet it still is! The system is still grinding away at it. And will continue to do so!"

We have another beer. I look at this guy who is able to talk about something so big so calmly. He tells me more. I hear that in his world they usually expect that a man won't be able to continue, that his nerve will fail, that he has to start fearing. And that he hasn't given in to that and won't.

We're in sync; the time has been rolling by, but somehow we speak the same language, nothing ever needing to be explained twice.

I decide to test my dream out again, since I have the opportunity, since I'm speaking with someone who has traveled the North Atlantic. Ships are ships and boats are boats, but the area is the same. I'm sure I can get some sort of opinion; I don't have anything to lose.

I tell about my plan, explaining my way of thinking, the essential points for the success of the operation. The correctness of the weather forecasts and their unerring interpretation, the right mastery of steering technique for safety and fuel economy, along with endurance and keeping my nerve. That's it in a nutshell, but there is an awful lot there.

The red beret listens attentively, thinks, knits his bushy eyebrows, and then quickly bangs out:

"Doggonit! That's exactly how to do it, and it wouldn't work any other way. Oh boy, what an idea!"

Harry comes back cheerily for a break, and my new friend starts to leave. I've enjoyed our meeting; half an hour just flew by. Next to the steering cabin he suddenly turns, claps me on the shoulders and says:

"Damn, what an idea, I tell ya', what a thing! It makes me want to go off with you!"

My friend is disappearing into the mass of people when I manage to call him back.

"Listen, it was bloody nice to talk. I wish you nothing but good winds. But who are you?"

He takes out a scrap of paper and writes. I give my own information and we part.

There is a small crush at the counter. A middle-aged woman appears in the middle of it, fingers a jar of spiced fish, and is already opening her bag. I am dumbstruck; all I see in front of me is a genteel blueness. She is sea-blue all over. Hat, earrings, eye makeup, scarf, jacket, gloves, dress, bag, and sleek boots. Lilac lipstick completes the creation. I'm in a public place, selling fish on my friend's slate, but I react verbally. I have to. The sight dazzles me, and I can already hear myself saying:

"My dear lady. I don't mean to intrude, but your blue makes my heart race."

The woman reacts with a start, quickly glancing at me. I see a clean-featured face and demure brightness. She withdraws farther off. I package salted herring and black bread, tell someone a recipe for spiced fish, collect money from another, and chat. At the same time I'm searching for the blue woman, the most beautiful thing I've seen today. She has disappeared.

In my hand is the piece of paper the red beret gave me, with a name and phone number. Fascinating, complex handwriting. It would certainly be nice to meet sometime, to sit longer, perhaps in front of some pints in a bar.

So Let's Build a Boat

The November morning is balmy, but there's a nasty day of sleet ahead. I park my car in Kellokoski in the yard of Juha Snell Ltd. I look, curious, at the simple factory hall. This is where they make Faster boats. I'm about to meet a fellow who understood what I was getting at after only a ten minute telephone conversation and just candidly observed that, "Sure, we can consider anything; come on down and we'll talk it over." For the first time I have a clear, real hope of getting my hands on a boat; there may be someone who understands and believes.

I enter, tense, my mind filled with all the "No, no" answers that have come over the past few years in so many different forms. The room is small and unpretentious: binders, papers, metal test parts, an office machine or two. A heavy-set man slightly younger than me encourages me to take a seat. This is clearly a shop engrossed in real work, not a showy playpen for yuppie kids who know how to talk the language of trends and market segments. I'm starting to feel at home; we get straight down to business.

We need about half an hour. I go over the basic details, tell about myself, my experience on the Arctic Ocean, and my ideas about what makes a real, seaworthy boat, as well as my opinions on how the project needs to be carried out to be possible. We have an easy time chatting; Juha mostly listens, nodding now and then. He asks a few clarifying questions and then calmly declares:

"So let's build a boat. A normal seven-meter Faster and leave the cabin off. You gain a little on the weight and can carry as much fuel as you need."

He doesn't whine about the hard times, about the difficulty of the job, about the demands of the Atlantic, about anything. He clearly has faith in his equipment. We go out in the yard and look over a frame like the one he mentioned. I kick some snow out of the way and look quietly at the bow of the boat. There it is! A real, proper bow. Strong, narrow enough to soften the buffeting of the sea, and substantial enough to raise the boat to the top of a wave in heavy seas, cross-swells, and hard handling.

I smile and sigh; this is the moment I've been waiting for. At first sight I'm on first-name terms with this hull; this will do, if I can drive. I've made tests, driving, comparing, analyzing these things so much; these are things I know by experience, and I'm not about to downplay that knowledge. I can simply see that this will work. I find myself off balance. This was the last real possibility, the boat is the best one possible, and I can have one to use when I need it. Something has shifted. Everything.

We arrange that I can have a month and a half. The decision to build the boat has to be made soon, in mid-January; this is going to be tight.

My old Mercedes almost flies on the way back along Tuusulantie. My adrenaline is pumping again. I mean, in all of Finland I found exactly one boat manufacturer who was ready to make this decision, quickly and calmly, trusting his equipment and me. He was grinning when he said, "To tell the truth, the project was clear enough after just talking on the phone, but I wanted to see the man himself too. I already knew your accomplishments."

At home in front of the fireplace I load my pipe and try to focus. Time really is short. In one and a half months I have to get the boat's equipment worked out, get all of the money for the project lined up, and find the right man to be the navigator. It's easy enough to want to go to Iceland—even in an open boat—but there are a lot of questions to answer before actually setting out.

I decide to start with the crew—who will be sitting next to me in the navigator's seat if I do go? I need an incredibly tenacious, open-minded, absolutely cool-headed, sufficiently multilingual man who has experienced the ocean, a highly competent route finder. In addition, he has to have a firm character to stand up to my egotism, a strong will, and brutal forthrightness in these matters. There aren't men like that hanging around on every street corner, and I can't actually ask anyone to go along with me anyway. After all, it's possible that in spite of everything, no one will be returning from this trip. I'm not going to lure anyone in, but who could I trust, who would be up to it, who would have the ability, and who would want to go along himself?

Absolutely

I can see our shadows moving ghost-like on the wall of the cabin; we are alone. My pipe has gone out again; I light it and I consider what he has told me. I have heard about the torture of the incessant questioning, the pressure and the games of the big systems in which the events themselves don't matter, but rather who's going to be getting millions and who's going to be paying millions.

I add some wood to the stove; we sit in silence. I look at the old seal rifle over his head.

"Yes, I'm ready. Completely ready," he says.

"Look, some things are emotional things, for some reason. I don't know. But if I'm going, I want to go with you; there's nothing unclear about that. I have until mid-January—that's when work starts on the *FinnFaster*, if it starts. And if it doesn't, I'll never speak another word about the trip."

I fill our rum glasses and we clink them simply. I like his style: no vain solemnity—he tells about his harsh experiences harshly, but hovering in the background is a strong personal philosophy; he has thought things over a lot, but he can stand to listen. He is obviously tenacious. Skill—I don't need to doubt his navigation or weather expertise, and his nerves will hold even when he's exhausted. His dissimilarity confuses me at times, but we still understand each other with a mere half a word. The best thing is the austerity, so far removed from the passenger saloon.

The red beret is on the bench. I don't think of it as a *prima donna* thing; it's just his cap. He also revealed a sort of professional secret: apparently the red beret and fatigues let him walk in peace on the side streets of Chicago.

After the boat question was solved, I thought intensely about the crew, about finding a suitable navigator. Above all I thought about who there was who was qualified and would want to go. You don't go dragging someone into a game like this; the right men volunteer themselves, in one way or another; that's the starting point. I remembered the red beret from the herring market and called him immediately.

"Matti Pulli."

"Pekka Piri here. We met in the market a couple of months ago. Any chance you still remember me?"

"Absolutely."

"Do you remember my idea, my plan?"

"Absolutely I remember."

"Are you still interested?"

"Absolutely."

The confidence in his voice was easy to trust. I summed things up. A sea captain knows how to navigate and read weather maps. The environment is familiar, and we wouldn't have to rely on only my language abilities.

The question mark I saw was adapting to a small boat without a bridge, a first mate, a chief engineer, cabins, or hot food. It's a totally different world from conditions aboard ship. My navigator would have to forget his stripes; they would only be ballast in an open boat. Would he be able to let them go?

We came to my fishing cabin to clarify these issues as quickly as possible. Being in sync with each other is just as important to each of us. We know that unfounded naïveté could turn tragically fatal for both of us. Boats can be incredibly confined places emotionally, and this isn't any old jaunt we're talking about.

"One more thing, Matti, now that we're here."
"Shoot."
"I'm a coast shipmaster with a charter boat captain's license. You're a sea captain with four stripes on your sleeve. But, Matti, but: I am the skipper of the *FinnFaster*. This isn't something to be arranged or negotiated. It's a fact. As a professional you know that it always has to be like this, but now after many years you're getting knocked a damn lot of rungs down the ladder. Small boats are my world. They're completely different from ships, and that's why things are this way. We work as a team, but I make the skipper's decisions. The sea captain I need with me is so emotionally hard-core that he can handle this. This is a matter of trust; either you trust me or you drop out."

"Sure as shootin'. I wouldn't dare go off in any tub where the skipper isn't the skipper. My stripes aren't weighing me down; they're way too faded already. And besides—my uniform jacket is probably still in the closet in my cabin somewhere on the bottom of the Atlantic. I haven't picked up a new one to replace it, since I don't really care for them much. This can give you a little insight into my soul. I think what you said about teamwork is right on. Each does what he's best at in this bucket, side-by-side. It's good that way, no problem."

"Well, yeah, this is important stuff. It all has to be firmly settled. We can't start yackety-yaking about this in the eye of some storm, and this isn't some sort of role playing exercise. It's a simple division of labor. And I'm not going to come playing at skipper on your ship, because I wouldn't be able to."

"Mmm . . . And I wouldn't let you, either."

Dealing with this delicate matter makes things easier. Now we're on a much more even keel than before. I appreciate Matti's easy attitude. I had thought this over a lot ahead of time and decided that if I saw even the smallest flicker, the slightest annoyed gesture or expression at that point, the jig would be up. I would get someone else. If a small boat heads off on a real trip across the Baltic, the North Sea, and the North Atlantic and there is anything simmering or smoldering on this side of things, all the elements of a fiasco would be in place, and at worst that could mean absolutely anything. Emotional strength is just as important as professional skill. A sea captain who can't fit into the role of navigator without being in command isn't emotionally strong enough for my boat. I'm relieved, Matti has that edge too.

"You have a nice looking cylinder head there above the door."

"Yeah, I picked that up off some shore rocks, cleaned off most of the rust, and slapped it up there. I think it fits here."

"It's a good-looking, genuine thing. People have been driving those around here for decades."

"Nestor was a man of progress. He got the village's first radio sometime during the war, was at the head of moving from sails to motors, and read the national newspaper; so Harry says. Someone in the village was a bit put out when he started getting the *Hufvudstadsbladet* instead of the Ålandstidningen so Nestor sort of tossed over his shoulder that he wanted to read something more in the paper than how many calves the neighbor's cow just had. It's Nestor's old head."

"Is he still alive?"

"Yeah, in Mariehamn in a rest home. He's a sort of good-natured scalawag. Elina often says that she's never seen anyone with such warm eyes. He likes a good drink, the old smuggler. Their oldest son, Hemming—he's first mate on the Baltic Star—pulled a good prank last summer. He came home on leave, got his dad from the old folks' home, and planted him in the front seat of his car.

Then he whipped open a bottle of whiskey and slapped it in the old man's hand. He drove his 80-year-old pop around Åland all day with Nestor taking swigs and singing out the window."

"Late in the evening he returned his father to the home as high as a kite and wondered why the staff was aghast. Old mother Vivi carped at Hemming a little, but understood completely that he had done it straight from his heart. Hemming sacrificed a day of his holiday and wanted his old man to have one day of vacation from the home too. And he knew his dad . . ."

The night wears on. Our stories fly from liquor running to Franciscan monks and the old votive ship at the church, from an ill-starred pilot to the fishermen of Nordkapp and the marriage of the Viking sea god Ægir. We both like stories.

Yes or No

I have to call tomorrow. Yes or no. Maybe or perhaps won't do. Juha Snell is waiting for my decision. You can't "maybe" build a boat—the aluminum sheets are either cut to shape or they aren't cut. There isn't any "maybe" about a welding seam.

A month and a half have passed since our meeting, and I've been working intensely on putting the project together. Just getting a boat isn't enough. We need motors, navigation equipment, radios, a ton of fuel, return transport from Iceland, survival suits, a big pile of expensive sea charts—the list is long.

A January storm is swirling snow around the light in the yard; I'm looking at the familiar view from my office window in Tapaninkylä. Some talk show is on the TV; Elina is watching and Arttu is in the shower following his soccer practice. Despite the seeming ordinariness of the evening, for me it signifies a crossroads, whose extreme branches lead to the stately grave of the Atlantic, the fulfillment my dream, or unbearable surrender.

I'm trying to track the swirling of the individual flakes. Is there some regularity in it, some law governing their paths? Is there some logic to the storms of my life? Why am I thinking about a journey to Iceland in an open boat on this January night? Right now, in the middle of this cold snow storm the most comfortable thing would be to add some wood to the fire and leave the frigid Atlantic as far as possible from my thoughts.

Would I be able to give up the whole idea? How could I, when it hasn't left me alone for three years? The hunger has only grown. I've done a lot of research, and I trust my work. I have strong experience, even though I haven't been on this route. I can see that the tension is coming from the fact that no one has made a similar journey. There aren't any examples, any models to follow. Has no one dared? Didn't anyone know how to do it? Or did it just not occur to anyone? I don't believe that last one. Someone must have thought of it. At least in Norway, right next to the Viking routes, it must have come to mind. For some reason, to me, a son of the relatively arid plains of Ostrobothnia, who didn't learn to swim until he was 11, the challenge is real and concrete. I have the best possible companion as a navigator—maybe Matti and I have the right to be the first! What do others' daring and skill have to do with us; this journey is ours from start to finish.

The thought of this trip fills some big hole in my soul. When I close my eyes and think of the rising of the silhouette of the shore over the horizon, I realize that I have to experience it; otherwise I will remain a cripple. It will happen now or never. Of course I'm going to go. It strikes me how funny this all is. What am I pondering; it's obvious! I'm going this spring. The time is now!

When I call Juha in the morning, I'm going to commit fully to this. He will assign one man to dedicate his time to building the *FinnFaster*, set aside a building spot in his factory hall, order the raw materials and supplies—a whole lot of concrete things that cost money are about to happen. He trusts me. I can't come mewling on May Day that, well, maybe I'm not going to go after all. . . . The decision has been made, the obvious decision. My sweaty dithering around is behind me, but it has been necessary. Without it, the decision wouldn't be solid.

On the evening news they tell about the Liberian flagged oil tanker Braer breaking up near the Shetland Islands, right along my route. I contemplate the accident, but it doesn't affect my decision; every trip is different. Besides: the small *FinnFaster* has its advantages too—at least it won't break up in the waves!

I've slept well; the snowfall has subsided by morning. I call right at eight o'clock.

"Snell."

"Hi, it's Pekka. Everything's all clear; start building the boat. I'm starting the engine on the seventeenth of May. Make a note of that; it isn't going to change."

"So that's the way it is, by God. I guess there's nothing for it. We'll do it. I'll have Timo Rokka start cutting the sheet metal by the end of the week."

The Blue Letter

The office corridor is clean, with fresh colors and sea-themed decorations on the walls. I find the secretary easily.

"Good morning. I'm Pekka Piri. I have an appointment with your boss."

"Welcome. Was your appointment about some sort of boat trip sponsorship?" She smiles.

"Yes indeed."

"Well, we'll see how it goes. Things like that are pretty rare around here."

"Yeah, well. This trip isn't exactly ordinary either."

"He's free now. You can go ahead and go in. Good luck!" She grins a little.

The director's room is spacious, with the spring light flooding in. He seems formal, clearing his throat and encouraging me to sit down.

"You had some sponsorship proposal; what kind of trip is it? I'm in a bit of a rush."

"Two middle-aged Finnish men traveling by open boat from Helsinki to Iceland in early summer. I'm hoping to test out your newest gear in my boat. At the same time you would get some visibility."

"We don't have a practice of . . . What, did you say an *open* boat? To Iceland? You can't be serious?!"

"You heard right, and I'm serious. I'm casting off on the 17th of May."

"Well . . . this is a somewhat of an exceptional situation. . . . You understand, of course . . . we have to be working on a firm foundation. We don't have any prior knowledge. . . . I . . . I . . . I'm sorry, but there is no way we can do something like this."

"I understand. It's too bad, but I guess that's it. Thank you for your interest anyway."

I gather my papers and make to leave. I can feel his disbelieving gaze on my back. He wasn't the first to shake in his boots for his own equipment.

I'm about to leave when he bursts out, still incredulous and partially baffled:

"All the way to Iceland? Why on earth would you go there? It's so far away, and we have such great water here, the whole Baltic and everything."

"Because I like the outer islands. Goodbye."

The secretary looks at me inquisitively. She looks nice. "Well?"

"It doesn't seem you practice this sort of thing. You were right. And my offer won't be repeated."

I'm already at the door, but I quickly turn back and whisper in her ear:

"Listen. You could do with a smarter boss."

The race to get the *FinnFaster* outfitted is fierce. The total value of the necessary equipment is high. I call, fax, write summaries, negotiate, introduce myself and my project over and over again. . . . I'm an adman, a salesman, an errand boy, a diplomat, an artist, and a fanatic. I won't give up; the boat has to be fitted out in departure condition by May 17th. It has to be; this train is moving down the rails so fast that half steps are a distant memory. I name my project The Call of the Sagas. It's a good name; it tells a lot.

I get an interesting snapshot of the reactions of Finnish companies to surprising proposals; the spectrum is wide. Extreme endeavors divide attitudes; decisions are always made by people. I experience the best moments with Bosse at Comarco and Sippe at Nores: "God, what an idea. We have to be in on this . . ." Kjell Holm at Furuno, an experienced sea captain himself, smiles behind his pipe: "You've taken on a great challenge. Congratulations! I have full faith in your strategy; it's extremely professional . . . have some more coffee . . . I'd like to tell you an interesting story from the Sea of Japan about ten years ago . . . hey, we could help you otherwise too. I'll send you weather faxes from here to the ports; I have connections in Scotland. . . ." They're like boys joining in a game, but based on a foundation of professionalism. The euphoria over each person who understands, over every positive response, over each new friend carries me over the discomforts and disappointments.

I find the blue letter in a stack of papers. The envelope is long and narrow, the paper expensive and elegantly emblazoned. I open it again; I like looking at it.

"... *The President would like to inform you that she will be happy to receive you after you arrive in Iceland ... please contact us at your earliest convenience upon arrival ... we wish you a warm welcome....*"

I am humbled and touched. A head of state I greatly respect wants to meet us after we arrive. I respect the tenacious people of Iceland. They have a deep culture, and they live on a lonely island in the middle of the ocean. They are the descendants of the Vikings and seafarers if anyone is. I know that their leader's letter will add to Matti's and my perseverance. Such invitations are wont to be obeyed.

Now, today, the 14th of April, is the most important anniversary of my life—I was born 48 years ago; I began to exist.

The morning was typical spring weather, alternating sleet and sun—just like the landscape of my soul. My crises have deepened the valleys and raised the hills, with everything in the game all the time. My life has no longer flowed through the holy, numbing trinity of office, living room, and bedroom, where conversations can be predicted a week in advance.

I've had to call into question almost everything, rethinking what I want, what my resources are. Who am I really? I don't have my old roles; I have to get by on my own, and it's rough. They don't teach you this in college.

That's probably why I'm outfitting this boat—I want to see who it will be that takes on the waves in the middle of the North Atlantic, what kind of man comes ashore at Höfn. And I want to live the joy; I'm sure of that on my own behalf. Yes, that wheel is going to stay steady in my hand.

I met my doctor in the morning at half past eight. I feel gratitude towards him; he has been very persistent with me, and at times my faith in life has depended on him. "Everybody can only take what they can take. That applies to everyone, including you," he said the first time we met. I've been forced to consider that often, and it has also helped me understand the ways lots of other people have of coping. As I'm leaving, I laugh and say to him that, "I might just stay in Iceland if I happen to meet some rosy *dottir*."

"That's the spirit," he answered with a smile.

During the day I met with OMC's Hasse Johansson. We'll be able to get the motor from Belgium in time, which was a relief. We'll be slapping a powerful 175-horsepower, 2,600cc V6 grinder on the stern of the *FinnFaster*. The best specialist I know, Timo over at Grönman, has promised to calculate the best propeller options. That will play a crucial role in both fuel economy and drivability, and therefore also in safety. Things are progressing.

In the evening I'm sitting by the fireplace in my usual spot. The fire warms my whiskey glass, and the smoky aroma smells nice. My birthday has actually been pretty good.

Elina hands me a piece of paper. It's her arbitrator's proposal for an agreement on the division of our property. I read over it quickly once and shout to her in the kitchen.

"Do you think this is good?"

"I think it's fair."

"Well, that's all there is to it then."

With that I scrawl my name on the bottom and hand the paper back. In my mind I marvel at how easy it was, dividing the material results of two people's efforts over 26 years.

Goodbye, Mom

A tired elderly woman lies in a hospital bed. I've driven a long way a couple of weeks before our departure to see her.

It's just the two of us, and the spring sun is trying to force its way through the gap between the blue curtains.

I look at the old woman. She clearly hasn't known me for a few years now, and hasn't even let me touch her. She is my mother.

It was she who gave birth to me long ago, nurtured me and struggled to raise me. From afar I hear a bedtime story read in a gentle voice, *Pessi and Illusia*. She lay in the bed with my sister in her right arm and me in the left. The warm voice read the fairy tale with gentle empathy. We were happy and safe there under that evening lamp.

I look at her delicate features, memories careening about in my consciousness. I came to her because I am setting out in a small boat farther than ever before. I try to talk, but she isn't able to follow along. I miss her deeply; I can see that she is already far away on her own journey and feel strongly that this meeting will be our last. Either her strength will soon fail her, or I might disappear myself in the Atlantic due to some miscalculation.

I take the weary hand in my own and stroke it gently. I feel a weak squeeze and realize like lightning that she isn't pulling away like before after all. I stroke her hand for a long time. This is good for us. This little thing is actually a big thing, actually the size of a life.

Mom turns and tries to sleep. I move quietly aside and look at her. Time stops for me. She is resting peacefully. After many years I got to stroke her; for a few fleeting seconds I had the opportunity to try to bring warmth to my tired mother.

I stand silently for a long while. My tears burst forth like a torrent; I cry deeply, silently. It comes from somewhere very far away, from childhood, from longing, from compassion. It also comes from all the joy and success and pressure and exhausting gauntlets run in my entire 48-year life. Mom has fallen asleep.

I walk for a moment in the bright spring of the hospital park; my mood is serene. I've said goodbye to her, and I got to be close for a moment. I got to stroke her tired hand. I'm happy that I exist; life had this moment in store for me too. Goodbye, Mom. I will continue my journey.

The sun has gone down, and the horizon is shining orange. The warm diesel of my car is ticking softly, and the kilometers to Helsinki are counting down. The radio is playing Jörgen Petersen's heaven-penetrating trumpet. I start looking for somewhere to fill up. My phone starts beeping. It's Matti. We talk a bit about the boat's safety arrangements—the placement of the life raft, protection for the emergency transmitter, and the food concentrate. The open expanse of Ostrobothnia is veiled in the fragrant mist of the spring evening.

Can You Dig It?

The may night is crisp. A white Mercedes taxi pulls up at the empty Viiskulma stop. The yellow top light switches on. The driver turns the radio down and thinks. As he thinks he glances at the odometer: . . . nearly half a million . . . in a base model Mercedes . . . a lot of it me driving . . . quite a trip . . . it would be a unique vacation, there's no way I'd ever end up going there otherwise. . . . Perttu has said quite a lot about it. . . .

The phone rings.

"Veku Pietilä."

"Perttu here. How's it hanging?"

"Just like usual. Things will be slowing down soon. Where are you?"

"On Syystie. Did my pop get a hold of you?"

"Yeah, he called just a little while ago."

"Did you get it settled?"

"Yep, we nailed it down. I'm going out and Anne and Jonna are coming along."

"That's damn good to hear. I'm sure it'll be a good trip. The Bergen road is good. We were there four years ago. It climbs over a kilometer."

"I've heard. Your dad has everything figured out. Put the combo trailer on the Viking Line to Stockholm, from there to Bergen over land, and from there to Iceland by ship. Back to

Stockholm with the boat on, and then they mean to return to Helsinki by water again. Everything's in place, and we'll be on the move in a couple of weeks. It's a four-wheel drive automatic, Ford, and the whole package on the way back will weigh six tons. Plenty of opportunity to put the pedal down. Can you dig it?"

"It's nice you got it worked out. A vacation'll do you good. Let's meet at the Puksu Esso station in a couple of hours when things calm down."

"Okay, I'll be there. . . . Looks like a customer . . . later."

A lone taxi accelerates along the West Highway towards Kivenlahti. The cargo is doing some groping in the back seat, stinking of whiskey. On the Lauttasaari Island bridge, Veku glances at Tiirasaari Island and thinks about communications . . . the SUV has a fixed NMT-450 phone and a handheld version of the same. They have the same setup in the boat and a marine radio. Pekka said he had arranged with someone in Helsinki to be a safety contact. Very interesting . . .

Muru

"... Finnair domestic flight AY ... has landed. Gate C2 B." The announcement goes through me like an electrical shock. She's coming soon.

The journey, our departure tomorrow, all the frantic preparation is somewhere far away. Now there is only this moment. We have an hour before her meeting. Right now it is more important than anything. I wait for her feverishly.

Am I really at the right gate? We have to be able to meet; we can't miss each other! I calm myself down; I know from experience that if we've agreed on it, we will meet. We read each other's thoughts and would come up with the same meeting place at the same time.

I watch the people in the bustling Monday morning terminal: people leaving, people arriving, people waiting, and hurried employees. I like the commotion of it all, trying to guess at the reasons for and feelings associated with each trip, destinies crisscrossing in the tidy hall. There an elderly gentleman with his luggage, here a young woman with a shoulder bag, a middle-aged lady next to me waiting. How has life been treating them? What are their calls of the sagas? Do they intend to follow them?

A petite woman descends the escalator and heads straight toward me. It's Muru! I recognize the energetic bearing and rhythmic, swaying gait from a distance. She smiles, a little embarrassed. I don't see anything else. Here she comes . . .

"To think that you're there next to me," she says as we drive along Tuusulantie. She touches my hand.

"Muru, darling. Where have you been?"

"Far away," she sighs. "Very far away." I can see that she is close to tears.

The trees along Mäkelänkatu are greening; the air is crisp and springlike, and a streetcar rattles cheerfully through the park. The young driver of the taxi ahead of us is dangling a hand casually out the window. We don't talk much. We're both happily embarrassed; we haven't seen each other in a year and a half. I can sense her familiar scent; her hair is in a little ponytail. I like it. She's an adult woman and a little girl at the same time. My mind is boiling; I want to kiss her this very instant.

"The stop!" she exclaims.

I turn quickly into the bus stop; again our thoughts met. I rejoice at her uncomplicated closeness. Muru plunges at my neck and kisses my cheeks, my mouth, and my forehead, mussing my hair. We laugh and hug. There are morning travelers standing at the stop, and we only continue on our way once a bus honks repeatedly behind us.

The brakes of the Merc screech and the nose dips violently. I park quickly in an obvious ticket spot; we only have an hour. The bright light of the May morning sparkles and the terns are shrieking in the bay of the South Harbor. We walk slowly to the Katajanokka Congress Hotel café. Our hands meet.

My coffee cup cools; I don't have time to drink it. After so long, all I can do is look at Muru. I see the vivid face, smiles and seriousness alternating in turns. The large, sensitive eyes look directly into mine and mischievous laughter interrupts our quiet conversation now and then.

I have suffered from her absence, the reason for which I did not know. I've tried to forget her but haven't been able to. Her lightness and encouragement have been constantly present, and to me her laughter is true female laughter.

A long time ago on a snowy sidewalk I started telling her about my stresses, my mistakes, and our whole family nightmare. She listened silently, not responding with a single phrase, just understanding and being present. She sent me poems, which seemed like they had been written for me alone. I took heart and experienced joy in the midst of the chaos.

Later the mail brought a beautiful card to my desk which read: "Fisherman, lower your net. Find the strength to try one more time." I turned on the red light on my door and looked outside silently. Through the window I saw the factory buildings, but I was close to her, and she did not scorn my exhaustion, because she knew so much.

She remembered and truly encouraged me, providing new sparks for my dying belief in the future.

Tomorrow evening I'm departing on the journey of my life, and she is stroking my hand. The whole gauntlet of the years rushes through my mind; tears flow onto my cheeks, and I look at her. I am so tired. These six months have contained some five hundred meetings, memoranda, and appointments, of which not one has been irrelevant. Everyone has been about survival. In recent weeks I've accumulated a sleep deficit; I've only had time to deal with things that immediately affected getting us into motion. This whole intense pressure is released now, because Muru is near.

"Just think, Pekka: it's true. You're leaving tomorrow. You are going to do it. You're that kind of a man."

"Muru . . . the most important thing is that we met . . . that we're alive."

I leave her in Kruununhaka and continue on, tires squealing on the curves, through the Market Square to the Estonia Basin dock. The newspaper men want their pictures before we leave. I'm just on schedule. I turn "Besame mucho" up all the way on the radio and roll the window down, feeling alive.

How about Going to Harmaja First?

The seventeenth of May is sunny but cold. A chilly northeasterly wind has been blowing for several days and the weather men are predicting that the same pattern will be holding for a good while.

The *FinnFaster* is floating in the Estonia Basin dock, shiny new, the logos of her courageous supporters on her sides and fully fueled. The dock is buzzing with a few hundred people there to see us off. There are speeches and ceremonies and the brass quintet from the Guards Band plays convivially. We're leaving in a couple of hours. In the middle of the commotion, Matti tugs on my sleeve:

"The propeller came, I put it in the port box."

"Darn good thing it made it; the dudes at OMC are sharp. We'll switch it somewhere out of sight. Let's not start fiddling around with things here."

"Hell no, not here. This is quite the circus."

"I promised our sponsors visibility. Let's try to stick it out a while longer. At seven we'll blow. Right on the dot, just like I've been saying."

"No worries. Just wanna get moving." I retire from the hullabaloo to the driver's seat for a moment and look over where I will be working. There they are, all our important devices that are supposed to help us make it far beyond the Baltic, the North Sea, and the North Atlantic. The unknown before us excites me, even though I feel so tired. The preparations have kept me so busy; on many nights I've only had time to sleep a few hours.

Matti's station is simple: the radar screen, the small satellite navigation display and working space for sea charts. With their help and his experience, he will have to be able to give me a precise course heading in any conditions—high seas, fog, darkness—for many weeks. We can't afford wandering about.

I have myself, straight in front of me on the desk, a magnetic compass, depth and speed indicators, an electronic compass, and a GPS. In addition, on the dashboard is a tachometer, gauges for the three fuel tanks, the engine hourmeter, a trim gauge, light switches, and trim tab adjustments—technology that doesn't do anything itself, but only helps when used properly and in working order. I'm counting on my instruments; only the best equipment was good enough for me, and I got it. My fingers are itching; finally I get to ride a real horse, for once there will be enough road even for me! Quick to the cabin on Kökar to sleep, and from there forward; I'm impatient.

Ceremonies, congratulations, and long looks into eyes. We know that a lot of people are afraid for us. We have given our explanations; we believe in what we're doing and we aren't going to sneak away. I remember my letter. I wrote it for my family in the early hours of the morning, as a precaution. ". . . if I disappear, I disappear a happy man . . ."

"Hi, Pekka. It's almost time to go. Magic time!"

The cheerful men from the Pouttu meat company lift the big crate of sausage onto the forecastle.

"Howdy, howdy. Ah, the grub has arrived. Put that there in the right side box in the bow. That's where the rations go."

The fellows' mood is contagious. I think with satisfaction of that forty pounds of the best salami. It will keep, making us self-sufficient. We'll always have something to chew on when hunger hits us.

Anneli and Kalervo meet each other as agreed on the restaurant terrace.

"Hello, Kale. Their departure is quite the carnival."

"Why not? They're setting off in style. Nice horn playing, hee hee. How are you doing?"

"Have you met Matti before?"

"No, but Pekka has told me plenty. Must be an experienced guy."

"Kale, listen. I'm really worried. Pekka is dead tired. I can see it. I know. And they are so frightfully different. Pekka is always so hard-edged about everything, and Matti just follows behind. What's going to come of that? I'm sort of anxious."

"That train already left the station. It can't be stopped, and do you think that Pekka would stop?"

"Like hell he would. Damn it. Hey, there's Jukka! Jukka, over here!"

I adjust the backrest of my seat and tune the gas suspension to match my weight. A familiar dark figure is waving from the pier, Pelle Krause from Maritim.

"Hey, Piri. Here are the rest of your maps."

"Is this all of them now?"

"I didn't get the one for Stavanger to Bergen, but I can order it to be waiting for you. You are stopping over in Stavanger, right?"

"Thanks a lot, but skip the order. I'll get them in Norway easily enough. I can handle it. We might not stop there. You never can know for sure. Thanks anyway."

We go to their Eteläranta shop to make the payment and sign for the valuable package.

"I'll use my Visa."

"No credit card. It's no good."

"Well then my debit card then."

"Well, see . . . I can only give them to you for cash."

"Only cash? What the hell?"

In the middle of all our rush, I walk to an ATM and withdraw the necessary thousands. Fortunately, there's still credit in the account. I return to the store and handle the payment. Pelle is a bit embarrassed. I get the receipt. At the door I start laughing and, as I leave, throw back:

"Yeah, I guess that's why cash is king. You're afraid the payer isn't going to come back, ha ha, that you'd get stuck with a bad debt. Listen up, Pelle, you're going to be selling me plenty of charts yet! Have a nice summer; we'll be off shortly."

The horns blast out "Anchors aweigh"—finally. The dock recedes. Another look and a wave. I train my eyes on my boys, laughing, on Elina farther back, quiet, on my friends, acquaintances, guests, supporters. I have finally started my journey. A wake behind us in the sea water, scenery changing. I make for Gustav's Sword and grin at Matti:

"How about going to Harmaja first to change the prop?"

The FinnFaster sweeps out of sight behind Valkosaari Island, flags fluttering. Anneli, Jukka and Kalervo sit for a moment more on the terrace, in the shade. The crowd on the dock thins. Lost in thought, Jukka looks at the people, enthusiastically buzzing, shaking heads, staring quietly into the distance. He is annoyed:
"Damn, they managed to get away so fast after all!"
"How so?"
"I didn't have any time to give Pekka this bottle of whiskey."
Anneli's voice quavers:
"Oh, Jukka! Let's hope . . . that you'll still be able to . . . someday."

After the Night, Morning

Chief Pilot Ingo Nordström is content in his profession. A week at the station and a week at home on the outskirts of Porvoo in Vålax is a pretty good rhythm. The marine environment and his office on the open sea are familiar; he is a third generation seafarer and there is a decent, professional atmosphere at the station with the pilots, cutter drivers, and all the rest of the gang.

The *FinnFaster* progresses slowly towards Harmaja. The familiar lighthouse tower is outlined against the bright evening sky. Many seafarers have kept lookout for this important navigation marker in the waters off Helsinki, and the bright flash of the mirror has offered many a feeling of relief when it finally appeared to their exhausted lookouts. I call the pilot station.

"Harmaja, *FinnFaster*."

The radio is silent.

"Harmaja, *FinnFaster*."

Still no answer. I wonder to myself why Harmaja isn't answering on the one-six, but I don't wait around wondering, I just drive in. I swing the boat behind the protection of the pilot station's breakwater. We have work to do. We need to quickly get on our way to Kökar to rest. I feel a little absent. That isn't any good at sea. An oppressive feeling washes over me momentarily.

The bright evening weather is cool. The sea will be radiating its cold for a long time yet. Nordström is walking in front of the pilot station; the day has been busy, as normal.

'... Sea Admiral, Dutch container ship ... West Harbor ... went easily. ... Maxim Gorky out from Katajanokka, old and heavy, lots of passengers ... gusty sea breeze off of Kruununhaka ... tried to sheer to the right ... helm over to the left, then went straight ... it was tight, made you think how to turn in Gustav's Sword ...'

Nordström squints at the tip of the breakwater, the setting sun shining in his eyes. Is there something there?

Keijo Manner, the cutter driver, has come out.

"What're you looking at, Ingo?"

"There's something at the breakwater. Let's go have a look."

Nordström and Manner walk along the long, rocky breakwater. An open aluminum boat is anchored at the tip and two men in red suits are bustling around at the stern, up to their waists in the water.

"Hey, that can't be them, can it? It was in the paper. All the way to Iceland ... could they be having trouble already ... ?"

"It is them. Let's go, quick ..."

"I called you but didn't get an answer. We just came to change out our prop. We'll be off in a jiff."

"Oh, no worries. Take your time. What boat is this? It looks like quite the pistol."

"A Snell seven-meter Faster. The motor is a 175 Evinrude."

"Ah. How much fuel can you carry?"

"Eight hundred."

"You must not run that dry."

"Nope, nope. We've tried to calculate it."

Nordström and Manner ask a lot of questions, and we chat congenially. In the questions I hear professionalism and respect for the sea. They aren't picking at our plan; we're just considering it together. Nordström's weathered face spreads into a grin:

"There's one thing in your route I've wondered about."

"What's that?"

"After the North Sea . . . bugger turning right. Go left, south where it's warm. . . ."

"Here's the thing, though . . . the sagas aren't down that way."

"No, no they aren't . . . and everyone follows his own course. . . . That's how it is. . . ."

The propeller is in place, the nut tightened, and the cotter pin inserted. We gather the tools in their bag. Nordström and Manner help with the ropes.

"We're ready. Thanks for the use of the shore. Would you shove our bow out a little?"

The men watch the boat receding to the southwest. It accelerates quickly south of Gråskärsbådan. Splashes of foam leap from the sides and the strong wind punishes the big Finnish flag high up on the radar arch. The FinnFaster has quickly risen up to full plane, and the men in their survival suits in the cockpit gradually become two red points far off on the Gulf of Finland.

They slowly return to the station. Nordström, starts to laugh, shaking his head.

"No. Hell no. We've seen a bit of everything here, but there aren't many who're going to believe this!"

Manner opens the station door, and the warmth from inside washes over them. He stops halfway in and looks at Nordström.

"What do you think, Ingo."

"It's a big deal. I believe in them . . . actually, it makes me a bit jealous. . . ."

"Same here . . . they're already a ways out, pushing west. . . ."

We devour the miles to Kökar. It's a chilly night, and the glow of the sunset has painted the sky violet. I let the boat to fly, getting up to nearly thirty knots. Matti is navigating next to me, calculating out loud. We don't really talk; we're free, but the circus of our departure and the blistering whirlwind of the last weeks continue in our subconscious.

The departure dock comes to mind. I swallow. So many close friends, so many generous handshakes, even from complete strangers. So many people understood why we were going. Dad came too. I understand his concern, when I put myself in the same situation.

Hanko Peninsula has fallen far to the right, and soon we are approaching Utö Island. The exhaustion is weighing on me, the compass growing dim in my eyes. We are silent, trying to settle in completely to our new life situation. I make a decision that Matti and I won't change places after all. I had first thought to sleep before Kökar, letting Matti drive, but I realize that it won't work. We aren't going to change places even once on our journey. In order to succeed we will need all of our know-how before Höfn. The skipper drives and the navigator navigates, always. Each does what he does best. Since both have their own work station the whole time, we won't have any dithering, hassling, or vague sharing of duties. We make a clear policy that we will only switch if we have to. This suits Matti too.

Finally, on the right Kökarören's familiar, dark shape rises from the horizon. We curve to the north into the blackish islets and early morning mist. I know the route by heart, and as the morning lightens we drive slowly into the sleeping, idyllic port of Karlby. I feel disappointed. I had imagined this moment being festive, but oppressive feelings have crept into my mind instead.

The rising sun illuminates the pink curtain of Lasse and Anne's lightly colored guest room. I look at it, weighed down. My brain is an aching anthill; there are too many questions. I know that I'm dead tired, but I can't sleep. Why did the fuel consumption seem too large? Why did location finding take so long off of Jussarö? Why didn't Harmaja respond? Why does Matti's approach to things feel so completely strange? Why didn't he get familiar with his instruments in advance? Höfn is distantly unreal; it has retreated somewhere behind a mountain of oppressive problems, smiling spitefully. I am completely alone and incapable of staying one more moment here inside. I escape outside, but the pain follows me. The nightmare is inside me. It's me—I alone am the skipper of the *Finn-Faster*, the father of the whole endeavor. I'm in deep shit.

An old, red open tractor with a trailer behind brakes on Karlby Road. An old man is driving, hunched over and wearing a black beret, without glancing to either side. Lasse's ducks have been quacking for a while now, Simons' cows are mooing for their morning milking, and someone is just starting the diesel engine of a fishing boat.

On the dock in the guest harbor, next to a powerful aluminum boat, a man in a blue sweater has been walking nervously throughout the early morning. Back and forth, back and forth . . . He is chain-smoking, has heavy bags under his eyes, and his face is strained and gray from fatigue. He always stops next to the boat, looks at it for a long time, and then continues his ceaseless walking.

Hell

My fist could crush the mobile phone. My cabin, my dear cabin I built with my own hands is there next to me. Torrskata Bay is shimmering in the morning sunshine. The weather is just the sort of budding summer that makes you want to pack a lunch and go on an expedition, and here I am smoldering on a bonfire I lit myself.

I'm walking around on the exposed bedrock, back and forth on the familiar rocks. So now I'm leaving you forever, am I? Just because I created this idiotic project, found a boat, a navigator, and all the equipment, planned, considered, worked things out and was willful . . . ?

There he's sitting in the boat. He has been silent all morning, and I've only answered his questions in half words. He hasn't asked anything more. We fall silent, and I know that something hard has crashed down between us.

I squeeze my phone, wanting to slam it down, just like everything else, straight onto the rock. I want to blow my whole plan to ashes to be whisked away in the wind and force everyone, every single person, to forget it forever. Not one word, ever! The phone beeps. I switch it off and flee like a wounded animal from the greedy journalists. I know the questions: How is the trip progressing? When will you be on the next leg? Is there anything in particular to report? How does it feel? And when was it you'll be in Iceland . . . ?

There would be plenty to tell, yes, the most delicious sorts of things for the likes of you, but I won't tell, I won't subject myself to ridicule. I shut my mouth.

I was serious. I had my purposes and my reasons. I didn't mislead anyone. You'd just mock me and measure how much space you would need for the sensational headlines. You don't want to come here to my bay really help me on my way; you just ask your questions and live off other people's nightmares!

Damn it to hell, I have the best possible boat at my disposal and top notch equipment installed by professionals. I'm certainly a qualified boat skipper and for a navigator I got a captain who has plied the Atlantic. And here I am walking in circles on the rocks at my cabin and those damn relays won't work and the guy just spreads his hands and says "that's life" and passes our sausage around the village boasting that "Pekka and me are going to Iceland even if we have to swim ass-first. . . . Who the hell is he? You can't plan anything with someone like that, least of all an ultra-long-range boat journey on the ocean, fucking fuck, I'm such an idiot . . .

The banquet in the visitor harbor restaurant was a nightmare. I couldn't tell the hosts to stop this farce right now, that I don't want to be your clown, I don't want to dive headlong to my destruction. Don't try to create a catastrophe! I didn't speak to anyone. I didn't listen. I wasn't full of enthusiasm after all. I just chewed my herring, thanked them, and went on my way.

My terrible alternatives have narrowed to two: if I call it off now, I'll be a ridiculous fugitive wherever I go. Here in my paradise is the last place I would dare show my face. If I go on I will perish, because the equipment and crew don't meet the requirements of the Atlantic. Evaluating that is my job. I'm the one who has to be able to make the right decision—I'm the skipper of the boat, and I can't deliberately set out on the ocean to destroy other people's property and two human lives.

Muru flashes in my mind. Oh Muru, you don't know where I am. If I abort, I will cease to be myself, and if I go on my life will end within a couple of weeks. This can't happen, but I don't see anything else; just when we came close again the day before yesterday after everything.

"No signal. There's something wrong with your radio."

This from Kökar's coastguard over my phone when we were trying out a test connection. In that moment the silence of the Harmaja pilot station and the strange loneliness of our nocturnal journey became clear. I had subconsciously wondered about the silence of the radio during our nighttime drive to Kökar. Not a hint of a call, not a single connection from the ether, but I was too tired to think about it, even though I always drive with the radio on. Its sounds are a part of my normal state of being at sea.

The crowd around on the guest dock murmured to each other: "Their radio is broken." I hated them. They aren't able to repair it, and I can't get a single sensible proposal. We're just an interesting performance routine that will fill the small village's need for something to watch for a few days and offer easy hindsight and saccharine laughter for years to come . . . "once there was this small boat that was heading off from Kökar all the way to Iceland with drums beating and all . . . !"

I've moved our departure time. "We're doing a little maintenance," I told the village fathers who were waiting for the ceremonies. In agony, I'm trying play for time. I'm walking around on the rocks and Matti is sitting in the boat. Out in the gulf swans are gliding and eider ducks are chattering. I call Juha.

"Serious problems."

"Ah."

"Radio mute."

"Huh, what's with it?"

"Even if I knew, there wouldn't be any help here. And that sea captain."

"Yeah?"

"Totally strange. Can't figure him out and don't trust him a bit. I don't even know if he knows how to navigate."

"Ah."

"I don't know what I'm going to do yet. I'm trying to think. I haven't been able to sleep for a long time."

"Yeah?"

"I'll be in touch. I can get the electronics in Turku or Mariehamn. Visby is damn far away and the weather will be bad here for a while yet. I also thought about coming back to Helsinki, full technical inspection and navigator exchange, but the press . . . I don't know yet. I can't even really think. I'll call."

"OK."

"Juha, I'm not complaining to you. I just have to talk, and this is your boat. I chose the man myself. My screw-up. I'll keep you informed. Later."

I can see Matti looking at me from the boat. I know that he's guessing I'm talking about him too, and that I'm not praising him. He's already gotten a taste of how tense I am. Amazement at his calmness flashes through my mind. He hasn't asked about who I was calling, about the schedule, about anything. And yet he sees that this is a bomb ready to go off, that the fuse is hissing loudly, that hard things are happening and we both have a lot at stake.

I glance at him again. Why isn't he fixing the radio, shouting to me that it's ready and that he's checked everything, that it's all working for sure now, and that now he gets this small boat thing, that this is completely different than a big ship, that here you have to stay busy and stretch and be precise with the tools. You have to keep everything in pocket-watch order. Now I know, Pekka . . . No, he sits in silence and looks at the rocks, the sea and the birds, but he also doesn't lash out at my rage. I don't understand . . .

I wasn't able to lay bare all of my anxiety to Juha, but he wouldn't have been able to help anyway. My underarms are soaking wet, and the telephone wants to slip from my sweaty palm.

It rings. I look at it, but don't do anything. It rings . . . a second time . . . a third time. I don't even move. I stare at the indicator light. Do the tabloids know yet? How? A fourth ring . . . a fifth . . .

"*FinnFaster*, Pekka Piri."

"What's up, traveling man? How are you doing?"

"Muru!!"

"Well of course. What's with the sighing? Aren't you going to give me a kiss?"

"Muru . . . you . . . Muru . . ."

"What's with you, Pekka?"

". . . Muru, this is hard to explain . . . this whole thing. The biggest mistake of my life. I don't trust him . . . this is a small boat . . . it's different than on a ship . . . ocean conditions. The radio is shit already and I don't know if anything else will work either. . . . We're going to die. . . . I don't want . . . and I can never come back again . . . not even here . . . ridiculed for the rest of my life . . . everywhere. . . . I'm tired. . . .

I hear Muru's silence, her presence. I see her near me. She asks quickly:

"When did you sleep last?"

"I don't know."

"Listen, Pekka. Listen to me now."

"Yeah, but . . ."

"Pekka, sit down right now on your ass, on that rock. Sit down right now. Are you sitting down comfortably, honey? Tell me."

"Yeah, yeah, yes, yes. What is it?"

"Listen. I know how much you've invested in this, and you know even better yourself. Think about it."

"Well, yeah, that's just the thing. . . ."

"You have to be able to rest, Pekka. Trust me. Not even you can go on like this forever. You can't make decisions when you're so tired, for your own sake. I can hear your situation and I understand you bloody well. No one can decide this better than you, no one. As long as you rest, darling. Promise me that you'll try. Do you promise?"

"I'll try, at some point."

No More Kidding Around

"I'll follow my own routes right up to Ören on my own. You try to get the bearing from there then."

I notice I'm talking to Matti as "you and I", not "we." There is a line between us, and it does not bode well. After my agonized wallowing, I found a compromise. I'll try my best to get us to Visby despite the bad weather without a radio, and hopefully we'll be able to get it working there, even though it isn't like one VHF guarantees anything in this.

There were the ceremonies on the pier, perhaps a little off-balanced because of the dithering about the schedule and my stress level.

Pages of Kökar residents added their names to the Norden Association's hollow baton we're carrying with us. The idea has been to bring it out at each of our ports of call as a message of Nordic unity. I'm indifferent to it. I don't have the energy to think about pieces of paper; I think, I think . . .

On the pier an old man came and pulled on my sleeve:

"Is it true that you will meet the President of Iceland, Vigdis Finnbogadóttir, when you arrive? I read something like that"

I was silent. The old man looked at me questioningly. My mind is too full of the broken marine radio, my doubts about the other instruments, Matti's unfamiliarity and uncertainty with his duties, a picture of the stormy Atlantic, an overturned boat, and two red-suited bodies dead from hypothermia being buffeted by the breakers. I was silent, looking into his eyes.

"That's the plan," I said evasively.

"I have a card for her. I made it myself." His eyes were shining.

Out of his breast pocket he dug a delicate card made of thin wood with a landscape, the president's name, and warm regards drawn by burning them into the wood. The signature at the bottom was Nils Volmar Sundberg, Kökar.

The old man looked at me brightly, trusting in me. What should I have said? Should I have told him that I'm a tired wreck, that I'm just trying to struggle my way to Visby and that I don't know a single thing farther than that? And that even Visby is a really long way off, even in good weather, and that my marine radio doesn't even work, and that my small, waterproof handheld radio is no replacement for a marine radio outfitted with a powerful antenna? And that neither can be heard on the ocean, even when they aren't broken.

That small, simple question added to my pressure. He was confident and excited, and all I had was doubt springing from unresolved problems. I saw Matti explaining our route enthusiastically. I couldn't promise the old man anything, so I answered like this:

"Listen. On the *FinnFaster*, the navigator handles all matters having to do with cards, charts, and maps."

I called Matti over and fled the scene, the elderly man's hope, praying in my mind that the card would make it to its destination someday, somehow, even if it was in the breast pocket of a corpse's survival suit.

The *FinnFaster* is rocking on the west side of Kökarören in a strong northerly wind. Before starting the motor I check the attachment of the zipper of my survival suit at my neck one more time. I only know one objective: Visby, about 160 nautical miles. I'm not thinking about anything else. I lean against the seat and headrest, looking at the sky. The clouds are racing hard to the south, and the sea is alive. The sun is sparkling, but the air is cold. I remember something. Once in December I went from here to meet Lasse's Vågen, nine miles to the south. Then there was fog and the sea was calm. I was just going to pick up a couple of fresh salmon. It was easy, pleasant . . . He gave me his position and I took a bearing. I was on his radar and he guided me in. He and Stigu took a break, we smoked some cigarettes, and I returned straight to the cabin by compass. Seems like I found a big sunken log on the way too . . .

"Are you ready?" I don't turn toward him.

"We can go," Matti answers. "Straight on."

"Did you mark the log reading?"
"One hundred ninety-eight and four, wasn't it?"
"It was, and the time is 1555. Put that down too."
"Let 'er rip."

Ören's wind generator is running hard and I raise the bow up, accelerate, and start searching the right attitude for the boat. A large gull rises lazily from in front of us. I watch the waves, the tachometer reading, and the speedometer. The frame is thumping, the sides throwing water far off to each side. The gull returns to wheel back to the islet.

The weather conditions worsen, the wave crests breaking, and the evening darkens. We haven't seen a single vessel, and we do not talk. The radio is as mute as a rock. We are alone. Driving eases my mood a bit; the sea demands increasing concentration. We dip deeply into the troughs and the oncoming slopes push on us powerfully. As we rise our speed fades, but we are making tolerable progress.

"Find me the highway."

". . . just a sec . . ."

On our satellite navigation system is a display like a road that makes it easy to follow a general course. I've never used something like it before, but I've been thinking it will make the long legs of our journey easier. Matti is pressing his buttons and a violent jolt sends his supplies flying into his face from the console. I keep the magnetic compass reading at the 213 degrees specified by Matti, but I don't slow down—the log shows slightly under twenty the whole time.

Night covers us. The dark Baltic rolls under us and past us, and we drive on in silence. I can feel the prop slipping under load. I can't maintain more than three thousand RPM on the uphill, which drops our speed to around 12 knots. The thought of changing propellers briefly flashes through my mind, but I let it be. I just drive. This speed is fine for now.

I am full of defiance. I'm not thinking of continuing on; I force it out of my mind. The only thing that is important is making it to Visby. That's what I decided and we have to get somewhere out of here!

"Do you want a banana?" Matti looks at me inquiringly.

"I'm not hungry."

I stare at the compass, handling the wheel with my right hand and playing the gas with my right, my thumb pressing the motor trim button. I'm driving completely alone. A bump sends Matti's T-square flying to the deck. I neither turn my head nor slow down one bit. He turns on his headlamp and crawls around looking for the T-square. I grind my teeth . . . dammit . . . let's his tools fly around . . . doesn't say anything . . . still no highway . . . on to Visby it is . . . we should be able to find it easily enough . . . fuel consumption . . . more than I calculated . . . I'm tired again . . .

"How much longer?"

High swells rise on the dark sea. I don't slow down.

"Just a sec . . . To Gotska Sandön Island about . . . twenty, a bit more."

Twenty. I count in my head. That makes a couple of more hours. The sea conditions aren't getting any easier. We could make it to Visby by three or four in the morning if we're lucky. We should probably report in with them on the emergency radio . . . I wonder if that breakwater is still there. It's been a long time, nearly twenty years . . . have to call Bosse on that shit radio to see if he knows someone . . . eyes don't want to stay open anymore . . .

"Have to loosen the fenders. Take the bow first."

On our left side is a rough concrete pier and behind it thick, reinforced steel netting. We took the first open slot. We didn't notify anyone; we just drove right into the harbor. Rain beats down on the pier. Matti ties the rope to the mooring post, and I start unrolling the bow canopy.

The canopy is almost in place, and I stretch toward the fastening pin on the right edge of the boat. Why can't I seem to reach? Why is it hard to stretch and keep my balance? I look at the pin for a long time. I just put this loop in there, like every time before. I stretch, but I have to stop. I can't do it. I know I'll fall into the dock basin if I continue. I, who have always been nimble in boats, enjoying balancing and never leaving anything unfinished. The fastener can stay open. I fight my way out of my survival suit and crawl into my sleeping bag. I don't have the energy to close the zipper. I'm shaking with cold and exhaustion, but I don't care.

Matti is standing on the dock. He sees a passenger car turning around into the harbor parking lot, as if checking the location of the *FinnFaster*. The lights fade into the dark and rain. His stomach is growling and he's so tired it's painful, but he wants to stand here and think for a moment.

What a performance! Fourteen hours of banging in a 16-meter-per-second north wind and not a word. What kind of a person is he? He just stared straight ahead and kept going as fast as he could. And he was obviously bossing me around. That was one hell of a race. Who is he really? Bowling around on the rocks like a tiger on the verge of exploding the whole time. What was that about? Taking locations was a trick in all of that; why didn't he slow down at all? He knows how to drive—I could see that right when we left Harmaja, but what's eating him? Is he trying to test me? That must be it. Crazy endurance—he hasn't slept for ages and yet we're here and in this weather. He's a completely different man from in the spring. He keeps asking about that highway all the time, even though I check it immediately every time I get a moment to look at it in peace . . .

Through my stupor I hear the trawler next to us starting its engine. The fishermen look for a moment at the silent *FinnFaster* floating there in the rain. Streams of water are running off the canopy, and the new aluminum shines down at the water line. The fenders creak and the ropes squeak. Snoring echoes from beneath the canopy.

I Just Want to Help

The Visby harbor is rain gray with clouds hanging low and a northerly wind pounding the rigging of the sailboats. Somewhere the radar reflector of a trawler is chattering. Terns are crouching on the ends of posts with their neck feathers up.

I sleep poorly, if at all. In my dreams I see crashing waves and breaking antennas, the tanks are empty and the maps wet shreds. Höfn is far away and the boat is adrift. The customs officer wakes me up with a start and with mixed feelings I go off in his car to the customs station to fill out the papers. Matti sleeps on.

Torrential rain pummels Visby all morning. Right there on that pier Anssi, skipper of the Jan Mayen, cut my hair with scissors 22 years ago. My son was six months old then and had just learned to walk holding on to supports. His grandma told us over the phone and it made Elina and me cry here so far away.

The customs officer is all keyed up about our trip:

"I read about it in the paper . . ."

"Yes, sure," I answer absentmindedly. I'm not really here. My thoughts are wrestling with fuel economy, instrument waterproofing, nightmares, the whole project. There's a lot out of joint, and I'm dead tired.

I don't answer my phone. Let them call. Lunches and invitations to coffee aren't going to drive us to Höfn. My shoes are squelching and I haven't washed myself in a couple of days.

"No tobacco, no liquor?" Not hardly. We have more at stake than some half-free bottle of booze, a lot more.

In the tavern someone is playing pinball. The scraps of Matti's and my pizza are growing cold on our plates. The pouring rain is banging on the windows and a chill wind is howling. We're supposed to leave in the evening, or should we leave at all? We're soaked through, we aren't talking much at all, and the mood is like a fiddle string.

"Miss, beer!" We look at each other for a moment in silence. Our bomb explodes. We have to set it off. I send the first salvo. We shoot hard, bellowing at each other. People cast sidelong glances at us, but we don't care. Many days' agony is released all at once. We shout, telling about all the things that are eating at us. Gauntlets are thrown down; we don't hold anything back. Fists are slammed on the table. The skinny guy at the pinball machine stares at us with his mouth hanging open and the bar hostess bustles about frantically behind her counter.

"You can't work like that!"

"Or you like that!"

"Don't you get it, damn you? What the hell!"

A long silence. I look at the wet, tired man before me. What did he say?

I just want to help.

That's what he said, in a quiet voice after so many days of bluster and steaming.

I'm silent. I look Matti in the eyes. He's wet, his face gray from fatigue, his eyes bloodshot. He says it clearly and expressionlessly. The tension in me begins to melt because I hear that important word. I heard that he wants to HELP. He wants to help me! He isn't an all-knowing, high and mighty sea captain who has traveled the world, he's just an ordinary guy my age, a tired man, who even after all my mistrust and bullying wants to help me and wants to continue the journey. I believe in the man who doesn't bluster. I don't believe in heroes, because there aren't any! Now I found something, now we can start together. Matti is my crew, I'm sure of that, at last. Joy and gratitude flood my mind. I regain some of my strength and my will begins to harden. I have a friend; he wants to help!

We are on the trail; I am resurrected to new life. Let the wind blow, let the rains come. The *FinnFaster* will set off tonight. Fill up the tanks, throw some grub in the bags, and off to Copenhagen!

I call Bosse and in a couple of hours a sweet pickup truck pulls up next to us on the dock. A grinning, elderly man begins inspecting our radio.

"The bug was in the antenna connection. It was broken, stretched too tight."

The phone rings in the boat; I go to answer it. I hear a shrill female voice.

"You're late! What's going on?"

"Where are you calling from, what are you talking about, and who are you?

"From here in Kalmar. You were supposed to be here yesterday. We've been waiting. We're angry! You were supposed to be here already yesterday. We read it in the paper! You have the message baton . . . We had a meeting and you were supposed to appear. . . . We thought . . ."

I count to ten extremely slowly, the corners of my mouth tightening. The azimuth before me doesn't even quiver. We were expected. The local chapter of the Norden Association is very put out at us. The radio has just been fixed, and I've gotten to rest a few hours. Matti is in the harbor office checking on the weather report. Where were we going anyway? To Iceland or to a meeting of old ladies in Kalmar. What is going to help us? Coffee klatch hosts or real action? I struggle for calm:

"Dear Lady. We are sorry, but we have had some problems."

"But we read the newspaper . . . !"

"Call the transit authority and ask for the schedule of boats to Iceland. This isn't a school bus!"

I hang up the phone. Shit. They're waiting for us here, they're waiting for us there. They've read about it in the paper . . . a meeting . . . so we were supposed to appear . . . because they had thought. . . . I think carefully. This isn't going to go this way! I call Hasse in Mariehamn immediately:

"Hi. It's Pekka. We made it to Visby. Just a few technical glitches. And we're tired. Our initial list of ports and times has made it into the papers. People are waiting around to throw us parties, but that's no good. We'll keep the promise we made, that the baton will go with us, but from here on out it's just a bow and a landing, that's it. Otherwise it will be Christmas before we get any farther than Norway. Give a brief announcement to the Western Scandinavian papers that there isn't any schedule, and that the *FinnFaster* is going to make its own way at its own speed. We'll handle Sweden here in Visby, Denmark in Copenhagen, and Norway in Bergen. Everything happens when we say it does—I'll try to give warning—but I'm not committing to anything. We aren't driving to any parties; we wouldn't get horse shit out of that. Weather and ports and refueling and maintenance and sleep and that's it. Matti also has to be able to calculate the legs in peace and get the weather reports. He has a hell of a lot to do on land, so we aren't going to start fiddling around with these lunches."

"It's good you called so soon. I understand perfectly. I'll take care of it. Pekka, listen. There's one more thing."

"Shoot."

"We're all completely behind you here in Åland. You've had troubles, I heard from Kökar, but everyone is with you, absolutely everyone. Tell Matti too."

"Thanks, Hasse . . . we will continue. . . . Sure, I'll tell him."

The Visby Norden Association group comes to the dock. I see bright umbrellas and clean trench coats. I declined the invitation for lunch. We don't have time. We're doing maintenance on the boat and making calculations. We invited them to the dock. It didn't seem completely self-explanatory to them, but they still came at the agreed time. We hear pretty speeches and try to respond, but in my innards I'm just thinking about what the baton really has to do with making it to our destination, with the *FinnFaster* really landing in distant Höfn so all these speeches don't just end up being empty liturgy.

The speeches end. Perfumed women offer us gifts and souvenirs. I get worried when I look at all their stuff: cans, bags, flagpoles, packages, envelopes . . . Goddamn it, why did they carry all this stuff down here? Who is loading my boat and on what authority? In Kökar we had a tough fight when we were getting the boat into traveling shape. Everything, absolutely everything out! I was harsh: we even counted the socks. Nothing extra! Nothing that isn't necessary. In a small, open boat it's better to have one wet wool sweater than five wet wool sweaters. This is what I declared on the shore by my cabin, and I know what I'm talking about. It is the foundation of our safety: all we have along is what we have to have and nothing else. In the lashing of the days and weeks anything superfluous becomes a dangerous, wet mush that no one can find anything in, and the chaos is complete.

We accept the gifts. Matti glances at me. I bite my lip, wondering frantically what to do. I can't refuse. They wouldn't understand. They would just be offended—we didn't come to lunch either. I begin to sense the solution. The *FinnFaster* is making its own way . . . ; that's what I just said to Hasse.

The pier has emptied. We are fully geared up to drive again, and I rev the engine for a minute to warm it up. Matti's calculations from the breakwater to Kalmar are clear. The afternoon is long and the pouring rain whips the deck of the boat. The wind has strengthened even more, but we are ready to start.

"I just want to help," Matti said in the tavern. That was exactly what he meant. I saw it in his eyes and heard it in his voice. Team, trust. We've found it. The *FinnFaster* has a crew.

I push the shifter forward, humming, ". . . in the old whale tavern . . . sits an old boatswain . . ." Matti slaps time with his sopping wet mittens, and I accelerate the boat up into a strong plane and rejoice in the rhythmic thumping of the hull. For a moment, a wake curving from behind the breakwater shows for a moment where we have gone.

The Joy of the Surf

The open sea begins off of Visby harbor immediately after the outer breakwater. A churning grayness surges from the right, from the north. The weather forecast seems to be right, the promised 15 to 18 meters per second from the north is coming true. In this wind, the motion of the sea has developed into a powerful spectacle stretching nearly one thousand kilometers from Kemi in the Bay of Bothnia.

We push out into the mix between the waves, into the deep valleys and crests. The next important target is the tip of Öland west-by-southwest, fifty nautical miles away. Forward from there the same distance is the harbor at Kalmar.

. . . Kalmar, Kalmar . . . The Kalmar Union . . . a skinny, insecure high school student leans tiredly on his desk thirty-three years ago. . . . The Kalmar Union . . . history class . . . something was agreed on there . . . good they were able to agree. . . . Jaska Tuulasvaara, "Teach" . . . you never droned on . . . once you praised me . . . I always remembered it; you encouraged me . . . now I'm steering a boat to Kalmar, with Matti . . . it feels good . . .

The pile of gifts on the bottom of the boat is clattering around revoltingly. I look at the flags. Is that a metal head on that pole? It's bounces nastily as we're battered around. In a big hit it could strike the windshield. It's dangerous!

"Take five degrees to the left."

"Five to the left." At the same time, I raise the bow a touch with the trim and play out the gas. We dip down into a gray valley. Luckily we have power in the back. A gust throws a bucketful of water onto the deck. The gifts float in the water for a second. The packages have been soaked through for a while now . . . damn pile of . . . that flagpole should be lashed down. . . .

A bright flash flares in my head: Where the hell are we dragging all this crap? Why? It's dangerous. Is it going to help us get to Höfn? Who decides this? I didn't ask for anything. I've been thinking about what we need for years. We're the ones out here doing this, alone . . . so we're late, are we? . . . no, never! I'm no one's clown!

"Matti, take a break!"

I look aft and starboard at the swells. They're much wider now; we've made several hours' progress from the harbor. We're almost to the middle. I'm going to do it; yes, I am. They're endangering our progress. They are a useless problem. They aren't going to help us get anywhere . . . !

"Break, what break? Now isn't the time!"

We have to shout over the noise of the sea, but we have the strength to shout.

"We're taking a breather now, just a short break!"

I put the engine in neutral and the smell of the exhaust washes over us. We rock, swaying in a broad movement. I glance back again. Now is good. I'm going to clear up this mess all at once; I'm going to solve a big problem.

I've climbed past the cockpit onto the bow deck and throw the flag pole as far as a can. It splashes into a slope and disappears immediately.

"Pekka, don't. You can't! Those were gifts!"

"Oh, so I can't, can I? Watch it fly, watch! Like this, watch!

Everything, every bundle! I'm not dicking around!"

Along with the unnecessary, dangerous gifts, I fling into the sea the whole surplus cargo of well-intentioned fussing, our role as cheap entertainment for others, the risk of paper mush clogging the drains, and the whole circus, whose clowns we were becoming . . . and everything else . . . fly, sink into the depths, and don't ever come back up . . . drown, all of you, forever . . . you goddamn vultures . . . stopping me living . . . as myself . . . if you try to resurface, my sons will crush you; they are strong. . . . I toss everything . . . the anxieties of childhood . . . the fears and taunting of youth . . . the humiliations . . . the grinding work as other people's pissing post, whose well-being no one ever asked about . . . the conniving bosses, the cowards . . . you sacrificed me; you were never motivated by courage, just the thought of your own position. . . . I toss all the rest . . . the refusals of my affection . . . no love . . . me just a camel, a beast of burden . . . for a long time . . . I throw and throw . . .

Every throw takes my breath away, but with every throw my soul is unburdened. I'm on my way to Höfn. I'm going there myself. Otherwise we'll never get there.

We'll do what we've promised, but we won't be burdened with anything else. We have enough work to do; the Baltic Sea is roiling around us, and we don't see anyone else out here. I don't feel guilt; I feel relief. I'm not a little boy next to a Christmas tree excitedly opening presents, and I'm not going to start lashing down extra flagpoles out here, when they're nothing but a nuisance.

The waves wash the packages towards Poland; I stride back to my driver's seat, out of breath. With that I accelerate up a slope and let the boat leap off it.

"Let's go, Matti . . . and fast; now we're moving! What course?"

"Two four eight . . . doggonit! If only I'd had a video camera! Take some sausage."

The late evening dims quickly, and the dark blanket of clouds is unbroken. The strong wind has intensified as predicted, at its peak up to 18 meters and above. Our spirits have only risen with the height of the waves; problems are something outside our ken.

I watch the behavior of the *FinnFaster*'s bow; it's working just like I guessed it would in the snow last winter in Kellokoski, even better. It takes the wave wall softly, but doesn't submerge even under hard hits. It just floats, not giving us a single jar's worth of water in from the bow, even though I'm driving really hard. This is a sea boat, and I get to be at its helm! The Evinrude grinding away behind me just pushes, exactly as I command, and doesn't run out of power even on hard up-hills. The propeller isn't slipping anymore; it's holding perfectly. Timo sure knew his numbers. He looked for the best solution that would take us the whole way without having to make any changes halfway as the waves, speeds, and salt concentrations change. Looks good, feels good.

I think about the nighttime drive from Kökar to Visby. We came straight into the harbor without wandering around at any point.

"Hey, Matti. How were you able to track our location last night the whole way?"

"It was no big deal, just a little harder sometimes because of the hammering, but it wasn't too bad. We just came straight on in."

"Why don't you ask for breaks sometimes? This doesn't have to be an express train. I've got the helm in any case. I always take breaks from positioning now and then; it is a lot easier since I don't have things to hold on to."

"Well, I thought you'd stop if you wanted to and we didn't talk about it otherwise. I just looked where we were going and a lot of times didn't say anything . . . you know how to hold a course . . ."

"Good God! And I thought that you would ask if you wanted at break, so I just let 'er rip!"

We start to talk about cooperation, about how we can get things to run more smoothly, how each of us can help the other, to make things easier. I know that we're drilling down to the heart of the matter better and more genuinely than in a single one of those seminars years ago in expensive forest hotels. We agree without question that the navigator will govern the breaks, since it's better for him to do his calculations in one place in peace than to rush and make mistakes. The driver may request a break when he needs one, a moment to rest and limber up.

We brainstorm and agree on a lot of things: Matti will always follow Greenwich Mean Time and I local time, we'll only keep chow in that bag, never anything else, the knife will always be in that loop right there. . . . precisely the little things that hold so much weight in a small boat.

"One thing, Pekka."

"Well?"

"I could handle that baton thing in the ports. We can still collect names if we have time, even if we aren't galloping off to luncheons. You were so frenzied up there on the bow and I heard what you were yelling when that flag pole flew."

"Well, what about it? I don't remember."

"You shouted with your beard quivering that if the baton and ceremonies got in the way of us getting to Höfn one more time, you were going to shove it up a whale's ass and we were going on without it . . . !"

"Oh, yeah, well . . . well, just so long as you handle it. That clears that up too. We'll keep it protected. It is a good thing; I really do value it or I wouldn't have taken it at all. It just made my blood boil when that old bag started carping at me . . . so we're late, are we? In Kalmar, when we get there, let's just sleep as long as we can, take on fuel, and then head off pronto towards Copenhagen."

"Sure, no problem. We didn't come here to set up house."

The phone rings; I can't hear the sound, but I see the indicator light.

"Hey, it's Heikki. Where are you at?"

"On the way to Kalmar. Be there sometime in the morning."

"How's it going?"

"Great. We've got about a 3-4 meter swell from the rear starboard. We're keeping up speed and surfing."

"Hell, yeah. That's a relief to hear."

"All our best to everybody there in Turku, Matti's yelling there next to me. Hey, it's coming on a little higher from behind. Got to put my hand on the gas and fiddle a bit. I'll call back at some point."

"We'll be waiting. I can hear the sound of the motor racing and the salt water flying. Good luck!"

Heikki hangs up and looks outside for a moment. Pekka and Matti are banging away out in the weather. Now is their turn. The roar was immense; the growling of the engine delivered to the telephone receiver through the satellites and copper wires, the wailing of the wind on the FinnFaster's flag ropes, and the avalanche of waves took him two years back to his trial by fire in the middle of the North Atlantic. A storm swept over them and Heikki's s/y Ikivihreä was in a tough spot, but they made it to Höfn.

Heikki thinks for a long time. A strong friendship has formed between him and the crew of the Call of the Sagas; in a way, he's there with them. Their age connects them too. We all let out our first cries during the same spring after the war.

Striving to reach Iceland in a ten-meter, open-sea sailboat and a powerful, seven-meter motor boat are different things and yet the same thing. Pekka is dependent on fuel; for a sailboat, the wind is enough. The sailboat was built for the open sea; in theory you could do a barrel role in it, but, being so slow, a sailboat is also more likely to end up in a storm. The FinnFaster is trying to find the best window of opportunity in the weather; Matti has to read the weather charts carefully, and Pekka has to be able to keep their speed up once they've set out on a leg, and they can't sleep until they reach port. A sailboat and a motorboat . . . and a ship—each one has its strengths, but the setting is the same, the sea is the same . . .

I enjoy driving; a feeling of relief and joy travels along with us. I get excited about testing the boat's limits;

I want to see what we can do with it. Little by little I increase the speed, carefully tracking the bow and the gauges . . . we don't go under, not by a long shot . . . one hundred rotations more, a little rise on the trim, still the same thing . . . the speed increases, everything is in control; it feels good.

I bump it up more, adjust a little bit more, but I don't take any risks. This package is full of wonders! I take the up-hills powerfully, without pussyfooting around; before the top I jerk the gas down low, so we won't jump, and on the downhill I accelerate to the limit, as much as I have time for, and trim the immersed portion of hull. The action of the bow gives me courage to screech down toward the troughs, even though the next slope is approaching ahead in high, solemn grandeur. Just before the base, I slow quickly, trim the engine relatively high up and throw the gas to full. I get the stern to curtsy and the bow to meet the oncoming slope high, again and again . . . the boat is like a part of my body and I am lithe again. I shout out the high speeds to Matti on the down-hills: "Two two four . . . now two five. Any higher? . . . nope . . . Hey, Matti, this is a record, two six two!"

Matti yells next to me:

"What a greyhound. This is incredible!"

I glance over at his face. Forty-eight years old, a sea captain who has sailed his whole adult life, he is taking pleasure in our flight like a little boy. I can't see even a hint of tension, even though it's a long trip from the lofty bridge of a ship down to the buffeting of a small boat below, at the level of the nocturnal waves.

Matti switches on the map light, raises his glasses to his eyes, checks the chart, and presses some buttons on the GPS. Then he turns off the light and bends down over the radar, looking for a long time, very carefully. The greenish light of the small screen illuminates his weathered face. Then he leans back calmly against his backrest and looks far off straight ahead.

I glance at the speed: thirteen. A forward wave is pushing us straight from behind; we've already been behind Öland for a long time. I raise both hands up to the center of the wheel, stretch them out straight for a second, breathe deeply, and lean back against the headrest. I can see the compass, the bow rails, and the surging waves, and look far out ahead.

The cloud cover is breaking up; the moon that has risen up ahead of us and to port is coloring the churning sea ahead of us silver. The foam shimmers as we rock, making steady progress. Far a-starboard a long, thin strip sits darkling, the shore of Southern Sweden. Before us on the horizon the smallest glimmer of light slowly appears. There sleeps Kalmar.

Vivi

The southern Baltic is made of scintillating crystal. We are already far past Kalmar towards Copenhagen. The mood is emphatically free; I'm easily able to maintain 28 knots. I could even run at full speed, 34, but I don't want to, I don't want to drive recklessly; I'm a grown man and the journey is long.

"Take a bit to the starboard, ten degrees. Watch out for that sand bank."

"OK, ten starboard," I say, acknowledging Matti's instruction.

Our sleep deficit is almost paid back; we slept a reasonably long time in Kalmar, tanked up and left greedy for our next leg. The pressure of four days of continuous wakefulness is lifting; the weather is pulling us forward, and we are joyously tumbling towards the Danish straits, completely new waters for me. We are really in motion. We are enjoying our freedom, rewinding the initial hell. We are open about how good we feel and talk a lot, even though long silences don't bother us. We're always belting out this or that snippet of song, together or alone; we're cutting powerfully through the Baltic. Now Matti's sonorous baritone breaks the silence:

". . . a wreetched waandering seeaman . . ." We laugh, and suddenly both sing at the same time:

"And brothers we ain't though we walk two by two . . ."

The miles pass, water flies, and Southern Sweden is far a-starboard. I luxuriate in the *FinnFaster*'s fine directional stability. The sun scorches our faces, and I light a pipe with two hands. I follow the azimuth, allowing the course to wobble five degrees to either side of the center line without touching the wheel. My record is seven minutes without direct control. I am satisfied; our boat moves firmly on its course.

A sunken log flashes by to the side—damn, that was close. We sharpen up our lookout; it's still a good hundred miles to Copenhagen. The four-millimeter hull will stand up to any concussion, and there are backup props for the motor, but we can't afford any collisions; at these mass velocities really bad things can happen.

I call Kökar. Vivi answers cheerfully, delighted at the call. She laughs happily when I tell her about our progress. I tell her openly about Muru and her call.

"Pekka, I am so happy for you. A person needs closeness."

"I'm good now."

"You're smacking your lips; are you eating something?"

"Matti and me are having a go at Lasse's smoked perch. Heads flying over our shoulders into the wake."

"Oh you young men!" She laughs airily, like a girl.

Vivi relates strongly to the excitement of long voyages. She is a wise woman who has seen many departures on the sea, a warm wife and mother. Now our pointing a small boat towards Iceland stirs her blood like before.

She remembers herring market trips with her Nestor, the weight of the big boat's tiller handle in stormy weather, and Nestor's mysterious departures all the way to the shores of Estonia and Latvia on dark autumn nights.

"I had a dream, Pekka. You returned; it was good to wake up. You are like my own son."

"I'll bring you something pretty back from Iceland . . . and say hi to Harry in Bergen. Take care, Vivi!"

I see her shawl-framed, smiling face, her body always moved to flowers, and her active hands, which over the decades have gathered bundles of leaves and twigs for the cows for the winter, milked, and rowed much, loaded herring into pots, woven rag rugs with long-tailed duck stewing on the stove. . . . With a heavy heart, she was also forced to bury her elementary schooler, her little Håkan, who had drowned in the bay by their home. She smiles often and says: "Life is like that."

"You are like my own son." I look at the compass with misting eyes. That was what she said, and she had shown it before. It moves me, because she means it, and it touches a sensitive place in me.

The sun sets on the clear horizon; the skyline is red. We take a break, find our location, and add wool under our survival suits. The silence of the sea is solemn again. We only rock one meter in the swell; there are still four hours left until the port. The moisture of the night sea condenses as an even film on the inner surface of the windshield; Matti has to read the radar carefully. The air is suddenly chill like in May; a Baltic Sea night is at hand.

"Take some gloves." Matti offers me a pair of neoprene diving gloves.

"Thanks, but I'll keep on driving with bare hands for a while. It's better for driving and my hands aren't cold yet."

"Ah. Well, I need them."

"Absolutely."

In the wee hours of the morning we reach Falsterbo Channel. The Baltic Sea is left behind and in front of us after the Danish straits await the great waters. The drama of the moment is real; the growl of the engine aft intensifies it. At the mouth, the turbulent current is strong; I have to keep the boat in heel with both hands, focusing.

Matti calls the channel on the radio, and we ask for permission to enter. He announces that a Finnish, seven-meter boat headed for Iceland is attempting to enter the waterway. The Swedish official asks wearily:

"Do we need to raise the bridge? Are you a freighter?"

Matti continues, face expressionless:

"No, we are the *FinnFaster* from Finland, seven meters. We're going to Iceland."

The radio goes silent. The deep stillness lasts an eternity; we're already starting to think we've lost the connection. Then we receive an emphatically dignified reply:

"*FinnFaster*. You are very welcome!" I push the boat into the turbulent current of the channel and try to take a photograph of the dark waterway framed by red and green lights which connects the Baltic sea to the Atlantic through the North Sea.

I feel like I'm driving into the most important channel of my life.

Early in the morning we weave through the colorful, mirror-like warren of Copenhagen and find a piece of pier suitable for a temporary mooring. Before sleeping we tie up the red and white Danish visitor's flag. Profoundly fatigued, we take guesses at what the half-asleep channel officer must have been thinking.

Despite the cold rain, at the edge of sleep I sense a warm softness in the early morning mist. Seen from Helsinki, we've moved far to the south, to the edge of Central Europe. From here on out we will only climb north-west, long and far.

I Couldn't Trust Anyone

Well-groomed people wander the street leisurely. The sunny Sunday afternoon has attracted families and tourists to stroll in Copenhagen's Nyhavn. We search for a place to eat; this is a time for a steak and large pint. Because it's a public holiday we can't get fuel for the boat; we wait calmly for the workweek to begin, which feels luxurious after nearly seven days of traveling.

We kick about and look around unhurriedly. Again I find myself observing people, their expressions and bearings, as if testing something; subconsciously I inspect every passer-by.

Matti buys some postcards at a kiosk; I lounge on a bench and watch the stream of people flow by, enjoying the rest.

"Look, Anton, this is a real sailors' street, just like they tell about in books!"

A neat German father in a sports jacket has bent down to his 10-year-old Anton and is pointing at something excitedly. Anton has on clean jeans, new trainers, and a light-colored sweatshirt; he has a few freckles and his hair is neatly combed. A mother with reddish hair waits for them farther off.

Anton looks, but doesn't understand. He probably sees the same thing I do: a clean pedestrian street for ordinary people: upstanding citizens, lots of families, children's strollers, and Sunday clothes. All wandering about unhurriedly. Anton's father's exuberance makes me laugh. As a boy he read an adventure book, and now he's showing Anton where it happened: "wild, wanton Nyhavn." He doesn't notice that here, today, they're offering souvenirs and food, and he doesn't remember that ships only stop briefly in ports.

Matti writes his cards. He has close friends around the world; the deck is thick.

"I'm sending this one to Canada. She's an old Indian tribal woman. I stayed with her for a week once."

I think of our nighttime entrance into Copenhagen. Matti was a navigation machine. The route was long and rambling; he was a clockmaker at work, quickly reading every signal amidst the big city lights, immediately giving me accurate readings. The hell of Kökar creeps into my mind; I haven't wanted to think about it. I can't even bear admitting to myself that I didn't trust him, that I even thought about replacing him. I don't talk about it; I can't. I listen to his stories about the postcard recipients and feel again how completely different we are, but I've started to like him in a new way, as a friend.

The Norden Association group visited our boat in the afternoon. They were fresh-faced young people for whom community with their neighboring nations was genuinely important. I liked them, their youth and enthusiasm, their sincere eyes. Under the canopy, as the rain pattered down, we told about our trip, and they wrote their names down for the baton; no one was in a hurry to get anywhere.

"But what is the deeper meaning of the Call of the Sagas?" the girl asked.

I was startled by her question. Deeper meaning? I had just been trying to explain that I've wandered at sea a lot, that I don't feel very old yet, that something in the voyages of the Vikings fascinates me and that I don't see any particularly strong borders between the Nordic countries, that I want to go somewhere far away now . . .

"Yes, but the deeper meaning?" she repeated.

Deeper and deeper! Who am I to explain it to these young geniuses? This twenty-something Danish girl was asking sincerely; she wanted to know. In school she has been taught to find out; she is part of a generation of world citizens, almost as old as my firstborn, and she wants to know the deeper significance of our journey.

"Look," I started, cautiously, instinctively brushing her hair lightly; I thought of her as my own child.

I continued, carefully weighing my words:

"When the years pile up and you see and experience all sorts of things . . . Sometimes you're not sure of anything really, like you were when you were young. . . . Sometimes you get tired . . . but still you want to live and feel . . . something, something important . . . and it bubbles to the surface . . . and then you want to leave. . . . Yes, I imagine it's the leaving that's the important thing . . ."

The girl looked straight into my eyes and thought. Then her face brightened quickly and she replied cheerfully:

"Leaving. That's a really good meaning!"

We have returned to our boat in the Langelinie marina. Rain clouds are building in the sky; I'm sitting alone on a bench by the shore in the small park. My mind is foaming with warm froth.

I look at the *FinnFaster*'s radar dome and antennas, but my thoughts are with Muru. She just called and said she was trying to arrange some time off when we arrive in Iceland; she wants to come! The idea of meeting her there is washing over me. I want to be alone here now, with her.

The evening darkens quickly. A man in a wind breaker suit jogs past our boat on the gravel path with his boxer. Suddenly he stops and comes back. He looks at the *FinnFaster* for a long time, his gaze shifting from the bow to the side. There he reads FINLAND—ÍSLAND, and then his eyes shift to the stern, the motors, and then rise to the floodlights and antennas on radar arch. He glances at his watch, but stands there a long time. Then he glances at the watch again and sets off jogging away briskly down the path.

The rain patters steadily on the canopy. We've just gotten in our sleeping bags and blown out the candles; it's after midnight. The sea chart for the Danish straits is on the bow bench; we were just looking at the route to Gothenburg for a moment. In the morning we will fill up the tank and leave as soon as we can. The forecast is promising steady conditions.

"Hey, Matti, I just thought of something exciting."

"What is it?"

"Back there in Nyhavn when we were walking around, I was looking at people's faces the whole time."

"And?"

"Now I understand what I was really looking for."

"Well, what?"

"I was trying to judge from their faces whether there was anyone I could consider a trip like this with. I was looking for fight and edge, but I didn't see it, at all. Not a single one I could trust, it felt like."

"Yeah, I have exactly the same feeling . . . and I don't think there'd have been many who'd have wanted to go either."

Damn It, It Won't Work

The night rain has eased to a warm drizzle. I've slept well and wake up alert for the brightening morning. It is the twenty-third of May, the name day for Lyydia and Lyyli; I don't know anyone with those names. I decide to give a report to YLE—the Finnish Broadcasting Company; that's what we agreed. I look at the marina park, the crews of a few guest boats about their morning chores, the bread-fed ducks next to the dock, and talk now and then into the phone. A recorder turning in Pasila, Helsinki, is storing my message; it will be played back later in the morning. I want to share my feelings with the listeners; maybe someone really is interested in precisely this morning moment in a small boat in Denmark under this canopy.

I start winding up my message: ". . . the coffee water is boiling and I'm going to start waking up Matti. We'll eat a bite and then go off looking for fuel; we're going to be following the west coast to Gothenburg . . ."

The surface of the Kattegat is calm and the hazy weather is clearing into scattered clouds. Our course is straight to Gothenburg, and we are able to cruise along at over twenty knots, even though I can feel the pull of the strong currents in the controls.

Refueling was hard work; Monday was a holiday here, and all the places were closed. We even did a loop between Helsingborg and Helsingør, and no one knew where we could get gas. I watched the boat's fuel tank gauges with concern; we didn't have much left anymore and looking was consuming gas too.

Finally we found a station that only worked automatically with bank notes. Luckily the village had a working ATM, and we got a thick stack of hundred crown notes, which modernity turned into 210 liters of gasoline. Matti pushed the notes in carefully, and I slowly filled the tank. I thought of the beaches in Helsinki—Puotila, Pohjoisranta, Merisatama . . . you can always find some there, and here they're living completely between seas for goodness' sake.

I correct our course and think about systems of holy days and working hours, the shared poker game of bishops, union bosses, and big-time industrialists. It starts to make me laugh. . . . I remember a farce of a negotiation about holidays and collective labor agreement terms, into which even the president finally waded . . . ayayay! My good mood bursts out as long, hard laughter.

"What are you laughing about?"

"Oh, this top-rated comedy show."

"Named what?"

"Oh, it doesn't have a name. It was damn-all long, almost twenty years, but I guess it could be called . . . wait. . . . yeah—its name could be 'The Jesuits' Polonaise.'"

"Is it still going on?"

"Yeah, I imagine so. The actors just change."

We've been making good progress for two and a half hours. The hull is bumping steadily, and I'm keeping Matti's course nearly due north. We sing bits of songs and trade stories. Sometimes I think of the edge of the North Sea waiting far ahead, the southwest corner of Norway. There the sea and air currents meet, the open winds of the Atlantic colliding with the high wall of mountains and the bottom depth varying in such a way as to easily make the waves break. The water there is heavy because of the salt and strikes with great force. On ancient maps they drew monsters there. Thinking things through ahead of time is conscious preparation; nothing should come as a surprise, and yet we know the unexpected will come anyway.

"Listen, Pekka."

"Yeah, what?"

"Actually, you should be offering me a bottle of booze!"

"How so?"

"We just passed the Kullaberg Peninsula on the right. Back in the old days, the younger sailor was supposed to offer the older a bottle right there."

"Ah, well . . . so that's how it is . . . but the difference is only a couple of months. You'll get a few shots in Höfn . . . if you can guide me there."

"Ah. Well, yeah, I guess I'll have to accept that you . . . you damn goal-oriented manager!"

The storytelling continues, one tale triggering another, and our stockpiles are significant. Sometimes we laugh so hard we cry; sometimes we are serious and silent. I notice an interesting thing: we don't tell a single joke—we tell true stories of events we've been involved in, and we both have a lot of them. The sun is warm, a gull or two glides with us for a moment, and out of sight on the right is the west coast of South Sweden.

"We'd left Chicago and almost come out from behind the breakwater, and I was catching some Z's when there was this god-awful thump. I flew out of my bunk and thought . . ."

"What the hell, Matti! Our motor is stopping!"

"Noo . . . What now? How is the fuel; it can't be out?"

The engine coughs a few times and we float in place. I turn over the engine—a couple of coughs and then silence. We are baffled. We check the handle positions on the tanks; they are extremely important and we have them written down in three different places. Everything is right—the forward tank is in use; we switched to it a moment earlier and it is full. I don't understand. The new, warm, six-banger engine shutting off like it was cut with a knife in the middle of a straight drive! I turn it over one more time without any result, not even a cough anymore. The small gear of the starter motor is accelerating the flywheel at full force; everything is as it should be—the sounds are right—only the comforting growl of the engine starting is lacking.

The echo only shows thirty-four meters; we lower the anchor and start to think about what to do. The thought of this situation on the North Sea or the northernmost Atlantic flashes through my mind. An east wind begins rising.

"Get me the number for the Halmstad Harbor office and call me as soon as you have it!"

I'm calling the Benefon receptionist; I have their best phone and remember all of the invaluable assistance I've received over the years from sharp, capable secretaries who knew how to get things done. We thought we should have that number just in case. In my mind I see the tabloid headlines: "*FinnFaster* laid up in Denmark" and we agree that we're going to try to get out of here calmly, by ourselves, and as quietly as possible.

"You can get the number yourself from the international information center. That's one of the nice things about our device. . . . First you press the function button and then . . ."

"Get the number for me and fast!" I raise my voice.

"It's really easy, just listen. First you push . . . and then . . . and then; you do have the manual there?"

"Goddammit, girl, knock it off with the function buttons! You get me that number right now and call me back. I don't have time to start fiddling with buttons and studying a manual. I have things to do and things to think about. And don't do anything for anyone else until you have this done. And you aren't going home until you've called me. Am I making myself clear?"

"Yeah, yeah, I'll call right back."

My anger is surfacing again. Damn it, like I'm going to start poking through manuals out here with the wind rising all the time and us having to start saving electricity and there is hardly any reception here in middle of the sea; it was a miracle that call went through.

We record our exact location and the time in the log book. The easterly wind is still rising. The phone rings; the secretary is calling.

"Here is your important number; do you need anything else?"

"Nothing else. Thank you for your help. We're having a bit of a rough go right now; we'll talk about it sometime. By the way, did some sailor sound off at you a minute ago?"

"There was a bit of an din."

"Ignore that. It was poor form; he's just having some problems right now. Greetings to your boss. The phones work well. I'm going to hang up now; we're using up our power."

We're deep into the evening; the engine is silent. In a boat being rocked by a strengthening wind in the middle of the Kattegat, two figures in red survival suits work without ceasing. They talk intensely, reasoning and considering and now and then trying the starter, but the engine remains silent. The fault lies in the fuel system, not the engine—that much is clear.

Ship traffic passes by some miles off. We could contact them by radio, but that would set the stage for a big time hassle. The lights of fishing vessels show up within sight now and then; we stay silent among the waves and weigh our options; we have to get out of here.

The power in the batteries has diminished, but there still is some. Because of the traffic we don't dare turn off the anchor light, which means the power will run out sometime after midnight; after that all that will be left is a hand-held radio that can carry a few miles and an emergency transmitter—nothing else. The wind is still rising, but it still feels like the anchor is holding.

"Bosse, get a message to Bosse!" Matti exclaims.

"Damn it, Bosse Boström, of course! He would be the best one now."

We plan the brief message ahead of time. As a precaution, in case it goes to an answering machine. We call the familiar number immediately, while we still have current.

In Sipoo an answering machine clicks on and records a reedy message:

"Bosse, *FinnFaster*. Twenty-third at twenty-three seventeen Scandinavian time. Location five-six two-five north and twelve zero-eight point five east. Motor not working, anchor still holding. Power running out. This is our last contact. Arrange quiet towing. Out."

We raise the bow canopy; all we can do is settle down and wait. We chew on salami and slurp water from the big canister. We've done what we could. Before we hit the sack, we let out another piece of anchor rope a little less than ten meters long—we don't want to go drifting towards the North Sea; we make fast what we can. Eventually we decide to just wait, to trust and to rest.

The nighttime wind has turned to the southeast and is increasing; the waves are growing steadily. The *FinnFaster* swings, held by her anchor. The bow sinks into the wave troughs and flounces up high on the peaks. Her movements and posture resemble a stallion imprisoned by its head collar, snorting, rearing up on its hind legs. The boat is dark; even the anchor light has gone out.

Under the canopy there is quiet chitchat: ". . . Once in the spring, Nestor was hunting seals out on Kökar's Karlbybadan and he lost his cylinder head . . . it was one of those under ten-meter open fishing boats. There was too much depth and so she ended up adrift and the north wind pushed her all the way down next to Gotland . . . there some little German cargo ship found her and towed her to Sandhamn in Sweden . . . then the authorities transferred her back home . . . it's been ages since then; for some reason the trip took more than a month . . . it's the same head you saw there over the cabin door . . ."

The stories under the canopy gradually thin out, and two even snores mix with the flapping of the flags and the whining of the ropes in the gusts of the Kattegat night.

A navigation officer on a cargo ship that has passed the Kullaberg Peninsula and is heading toward England via the Skagerrak notices a small point on his radar a few miles to one side. He inspects it for a moment. There aren't any buoys, there aren't any lights visible, and the point doesn't move. The significance of the point remains unknown to him; the ship continues on its course, and the point is slowly left behind.

Moon Shadow

The morning lightens as a faint, gray drizzle. Sleep in the violently rocking boat was not deep; we were forced to strain to stay on our benches. This complete stop, not knowing anything about the future, nags at us; all we can do is wait. Something will happen, sometime, no doubt, but we cannot know where the course of our lives is taking us now. The strong boat that had been progressing so swiftly now bounces mutely at the end of its rope, and we aren't singing or rejoicing as the miles pass by to the next waypoint. Höfn has receded into the distance again. I don't want to think about it; all I'm thinking about are the coming days and the problems we will face.

We weigh alternatives, trying to figure out the future, our actions, and to predict the possibilities. Bosse has our message—presumably, but we can't be absolutely sure. He might be at home, but he could also be traveling. If he receives it, he will have to operate from far off in Sipoo, but that doesn't worry us; that's why we directed our last watt to him and no one else.

The hold of the anchor is starting to reach its limits; the wind has already risen to around twelve meters. If we loose it, we will be adrift, and we won't able to find our location anymore, other than in a general way, since all of the navigation devices are cold. Drifting ashore is a risk to the equipment, and in rough conditions perfectly possible.

If Bosse's machine is broken, if our message isn't there after all and we don't break ourselves loose, they'll start looking for us at some point—that much is clear. And we will be found munching salami in the Danish straits with Iceland painted on the side of our boat. The papers will have their story, and the reporters will ignorantly write several versions about the problem we don't even know the cause of yet. I don't want that problem for my supporters.

We could get in touch with one of the ships that passes off in the distance now and then using our hand radio any time we want, but that would mean a big hassle and a lot of bullshit, so we stay silent. This is ultimately a question of what will happen first: Bosse coming through or the anchor giving way. Hurry, Bosse . . . you have a message waiting. You'll know what to do—you'll get going and make things happen once you get behind the radio . . . We'll start to slip soon, not many hours from now . . . the wind is rising the whole time . . . hurry, use all of your connections, fast . . . you'll find someone . . . we aren't going to give up; we're going to stay here and keep quiet . . . that's exactly why we called you . . .

Having a stop to all events and activities in the middle of the deserted, gray sea is thrilling even in its painfulness. We aren't in any real danger, but our hands are full of problems again. We have done what we can and now we are trying anticipate the future, organizing our tools and supplies after a long night of working. Our clothes stink of gasoline, our hands are black and oily, and we are tired and uncertain of the future, but we don't feel any great pressure. Concern, yes, pressure, no. The Estonia Basin is already so far behind us that we are living in this place and moment, our uncertain, bobbing situation here at the end of a rope 12.5 miles at 303 degrees from the Kullaberg Peninsula.

I sit in my driver's seat and look at the pale, cold gauges; only the magnetic compass works, the dial spinning in time with the rocking of the boat. We are dependent on technology and have known it since the beginning. The future weighs on our minds; we don't even know if the next time we dock it will be in Denmark or Sweden or if we'll drift into the North Sea. We don't know anything, but we try to anticipate things; we study maps of both shores of the Kattegat, the cities, the possible service points, our routes forward . . .

We're killing time, glancing at the rising peaks of the waves, and listening with our whole bodies to the anchor. We read the language of the sea and the ropes; we would feel any release immediately. We don't talk about it—it won't help—but we are listening to the same thing. The hours pass; we sit under the canopy and chat, whittling off hearty hunks of sausage. Technology and the forces of nature have created an unknown turn in the drama, and as the protagonists we can only wait. We know that sometime something will happen, and we will need our strength then. So we rest and eat salami and keep checking on the anchor.

We follow the direction of the rising wind and are ready to record down to the minute when we break away, knowing that the strong currents of the straits will influence our drifting; there's no way to calculate that with any certainty. I remember the many times when Lasse's salmon nets have drifted upwind, even in Kökar, and here the tides will also have an effect.

It is an unhurried early summer morning on the south side of the ancient, 1000-resident fishing community of Halmstad. The smells of sea and meadow melt into the cries of the terns as they whoosh along into the harbor bay separated by a long breakwater. The group of old men the same the world over are talking near the dock, weighing past and present cares. Pipes are smoked and stories make their rounds; this isn't about listening or telling, but just being there.

Coastguardsman Kjell Holmberg is spending his day off working on his boat. A nine-meter, solid, fishing-bodied craft with a forward cabin is floating in its normal spot, the easterly wind lifting and lowering the surface of the harbor in an even rhythm. The ropes and fenders creak now and then as if yawning in boredom.

Out the steering cabin window, Holmberg sees a neighbor climbing onto the foredeck; he looks like he's in a hurry and pushes his way inside.

"Kjell, I just got a call from Halmstad. There's some boat out on the water than needs to be towed. I have the info right here, the coordinates . . . something small, from Finland . . . been at anchor almost a day . . . the wind is rising . . ."

Holmberg draws a cross on his sea chart; the calculations are ready. He only thinks for a moment . . . a direct course from the breakwater opening to the target; the trip is about fourteen miles, which will make a couple of hours to get there on these seas . . . let's go have a look . . .

The powerful Perkins rumbles into life; Holmberg looses the ropes and backs out from the dock calmly. At the opening of the breakwater, he turns the radar on, increases revolutions and sets a course straight toward the cross he drew on his sea chart.

I'm on the bow, pissing. As usual, I'm scanning the gray horizon; I always do that at sea. The crests of the waves have been breaking for a long time now, and the streaks of foam are growing longer. I have to lean on the railing; I'm on my knees and my heavy weight is pushing the bow down, the splash of the crashing waves hitting my thighs, the largest up to my waist. My gaze circles the gray border of sky and sea; I see the restless sea, but I am looking carefully, because we are waiting for something, waiting for a solution. Bosse, or the anchor . . .

On the horizon, between the white breaking crests it's like there's a larger frothing. I stare at it; it wasn't there a minute ago. Am I just imagining it? I'm not. Its course stays steady, but it seems to be growing. It isn't a rock, and it isn't the sea's own foaming. I can sense the rhythm of the splashes; it isn't only coming from the sea—there is some other force involved. I can see the froth disappearing and then charging up again rhythmically and disappearing in the wind. Whatever it is, is being driven. It's coming from far off and not changing its course a hair; it's coming straight here!

I look on silently for a moment; I'm sure. In the distance a boat is struggling directly towards us, coming from the east, from Sweden. In my delight, I am selfish for a moment; I remain silent, breathing deeply. Here it comes! The froth grows and the course stays unchanged.

It's coming straight here, bow foaming; the empty waiting is over!

"Matti, put some deodorant in those pits, we've got company!"

". . . huh, what is it? Is there really something there, no fooling?"

"Yeah, the salvage team is coming. Damn Bosse, what a tough dude!"

We quickly begin to roll the bow canopy down, getting ready. We're going somewhere.

A stout, dark brown, high-bowed boat rolls powerfully up on our starboard side, fifty meters off. We discuss the maneuver over the radio. We hear that they have a fully charged battery for us, that they had been encouraged to take one as a precaution, but we let it be—transferring it to us would be dangerous in these seas, and it wouldn't solve our problems; we need more, a harbor.

The boat comes carefully closer. The throw line hits them on the first try; they attach a heavy tow rope to it and we pull it to ourselves. We make it fast and finally raise anchor. I look at the strong, metal grapple for a split second and thank it . . . you stuck it out too. . . . We drift downwind and the rope tightens. They invite us into the warm cabin; we thank them, but announce that we will be staying in the *FinnFaster*. We aren't going to leave her; we'll get along just fine, and don't even consider the cabin.

The towboat's engine grinds and a thick cloud bursts from its exhaust pipe. The boat has a beautiful shape—it resembles the northern Norwegian sjarks or English working boats, and on her side it reads *Moon Shadow*. In the log book, Matti records, "Towed by *Moon Shadow*. Towing began at 12.30."

After a couple of hours we are moored to a stout pier. The *Moon Shadow*'s diesel goes silent and an agile man comes out of the steering cabin, smiling.

"How was your trip on the end of the rope? The seas are pretty bad."

"Thanks, but there wasn't much to it. We just slept."

"You've got quite a ways to go. We heard and read on the side of the boat. Serious shit."

"Yeah, well . . . now we're having some trouble. There's something mysterious on the fuel side of things that we have to get figured out . . ."

"She came along nicely on the rope. We kind of wondered . . ."

"Yeah, she moves along well. . . . and can really . . . but where are we now?"

"This is Torekov. Welcome!"

People gather on the pier; they are friendly and helpful, but it's hard to understand their speech.

Shut Your Trap Already

We're on a walk on the edge of a beautiful, low peninsula. The environment is a scented idyll, but all I feel is the pressure, thinking and thinking . . . The important endpoint of our Scandinavian tour, Bergen, is an infinity away; between us and it is the difficult edge of the North Sea, and somewhere much, much farther off, an eternity away, are the Shetland Islands. From there another eternity away lie the Faroe Islands in the bosom of the Atlantic. Only from there does the actual journey begin, two new eternities, three hundred miles away over the North Atlantic to the ancient Viking harbor at the base of the Hornafjord, and here we are strolling along a damn path in southern Sweden already having been hauled at the end of a rope, without any information about the cause of the problem or whether we might see it again in the future.

We are between two worlds. Our wide world is a dream, reaching for a distant pier on the other side of a northern ocean. Our small world is full of unbearable problems, hoses, tools, technical details, concise calls on handheld phones, thinking, struggling with the Skåne dialect, and a bank machine that won't give any cash with my new credit card, as well as hunger, exhaustion, and wetness.

The stage for our small, problematic labyrinth is this ancient fishing port; the undulating, fragrant meadows smell intoxicating, telling the story of some six hundred years of human struggle on the edge of the sea. Even the terns scream more intensely here than anywhere else.

The delicate idyll and the depth of my problems are light years apart; the contradiction tears me apart, leaving me mute and deepening the dark wrinkles on my forehead. I would like to enjoy the fragrant timelessness of the early summer, but I am unable. I've never been able to rest amidst unsolved problems; I've rested really well very rarely.

"Look how bright the green is! The most beautiful sunset of my life! Look over there, Pekka, look!"

I suffer in silence.

"Birds, I love you. What a concert! Look at the buildings, this ancient harmony!"

I'm about to blow my top; I grind my teeth and strain to shut my ears.

"What sensitivity . . . look at the harbor. The boats have returned and now they're sleeping quietly . . ."

I lash out in rage:

"GODDAMMIT, SHUT YOUR TRAP ALREADY!"

Matti is startled. He stops and looks at me, eyes flashing. His face is as red as fire, and he's boiling with fury; he looks for the words, picks up speed, and then explodes, his whole body trembling:

"SHUT UP YOURSELF! I GET TO TALK SOMETIMES TOO . . . I'VE BEEN LISTENING TO YOU FOR A WEEK; YOU'RE ALWAYS TALKING! ON THE PHONE OR OFF . . . IT'S ALWAYS YOUR GODDAMN TURN TO TALK . . . YOU CAN BET I'M GOING TO TALK TOO . . . I SAID THAT I WANTED TO HELP, BUT I'M NOT GOING TO BE QUIET . . . AND THERE'S NOTHING YOU CAN DO ABOUT IT . . . !"

I'm dumbfounded. Matti draws a breath and wheezes, his face tight. I just saw a powerful drama; our cold motor was forgotten for a moment, and my brain received half a minute of compulsory rest. I also saw the power and will projected by his declaration, and it buoyed me up. I'm not alone after all, and my companion has shown he has an edge; I wouldn't be able to stand a yes-man. I thought I was alone again trying to solve everything, and it felt overwhelming, but Matti's anger stopped me in my tracks, encouraged me. There are two of us; we're sure to get things solved . . . My pressure is gone, and Matti is right; I'm domineering and I talk a lot, nowadays, when I can talk openly. For twenty years I couldn't really; I just had to fit words together, spouting generalities, speaking others' words, or keeping my mouth shut and playing my part.

"I . . . Matti . . . say just what you want . . . and I will too . . . We both have to talk . . . we just have to. . . . Just let it come; otherwise nothing will come of this. We're like night and day, but we're driving together here."

"I don't have any problem; I was just trying to cheer you up, since you were so down and, yeah, I understand the situation and that the extra costs are weighing you down on top of everything. That's what this means."

"Let's go to the village; the grill is probably still open. . . . In the morning we'll pick up the batteries from being charged."

The scent of the flat meadows and the peacefulness of the bluish haze fill my consciousness in an instant; I relax and enjoy them. It's right now that they are here; now is their moment. I want to record the smell of evening in Torekov in my soul.

Chance has thrown us into a fascinating place that would have remained forever unseen otherwise, and that day on the end of a rope, equipment cold in the middle of the drizzly Kattegat was a good, empowering day. In our boat we operate as a crew in complete unison, and we respect each other's expertise; on land we dare to talk about our differences, which aren't a problem anymore, but rather our strength, and neither of us is afraid of loud voices. There can't be any moping—we don't have room for it, and we don't need it. Every leg we travel together, every nighttime call at a port, and every solved puzzle has begun to unite us; we are an odd pair of horses yoked on a journey from one place to another place very far away. We live our own life in a pounding world seven by two and a half meters where water sprays over us, stories fly, and we sing loudly before and after the problems we encounter.

Thanks, Hans!

I look at Hans Fagerberg standing on the dock. He's smiling behind his bushy beard. It's six in the evening and our engine is idling smoothly. We're in our rescue suits, ready to set off for Gothenburg; the sea is waiting, calm and foggy, and we only have about a hundred miles ahead of us.

I feel light; I want to get at the next leg soon, and I'm itching to drive my boat. The foggy night ahead of us is no cause for worry; I'm completely confident of Matti at the radar and GPS, and I'm like a part of the *FinnFaster*—her handling in any conditions is something to be relished. Hans solved our problem. He sacrificed a day at his little dockyard for our troubles. He was working on the restoration of a beautiful wooden boat, and there was a real rush to get it done, but he understood us and so he came, grinning, to our aid. Even though the fully outfitted, aluminum *FinnFaster* is a far cry from his love, wooden boats, he wanted to help those in need and believed in our objective. His eyes shone as he said this; he wanted to be a part of this.

The problem wasn't in the boat or in the engine, but rather in the three-way valve that distributes fuel from our three separate tanks to the engine. It had started to leak, and wouldn't release the forward tank's fuel because the middle tank was empty. Hans worked it out systematically in his workshop and testing it in the boat. Fixing the problem would have required several days of waiting, which we didn't need. We would make it easily enough to Gothenburg without the bow tank, so we were ready to go again.

"Thanks, Hans! Thanks for taking the time and understanding."

"It was nice you stopped by. You'll make alright."

The ropes are off and I push the shifter forward; we begin to push into the opening in the breakwater. The *Moon Shadow* is left floating, stout in her berth.

We make good time towards Gothenburg in the late evening fog. We drive away from the unseen shore and talk for a long time about how warm Hans and Kjell were, understanding and helpful. Now and then you meet people like them.

It feels good to be in motion; Bergen is slowly growing closer, even though we do have that difficult edge of the North Sea between us and it. In many ways Bergen is my city, and I'm no longer aiming directly for Höfn. We can only make it there leg by leg.

After midnight we slowly weave through the misty light labyrinth of Gothenburg, looking from below up at the steel sides of the great ships. The metropolis hums around us as we tie our ropes to the dock Heikki had told us about. The fog is thickening.

The building site of the new opera house next to us inspires Matti to tell a story before we fall asleep. Once in Amsterdam he had gone with the radio operator to a symphony concert being conducted by Paavo Berglund, a Finn. Over the furious applause of the audience had come the rhythmic cry of two male Finnish voices from the balcony: "Way to go Pate, way to go! Way to go Pate, way to go!" They weren't the least bit ashamed; they were proud.

Before I fall asleep I hear the sound of a car siren. I think of my many countrymen who earn their bread here and think about how there is a direct shipping line from here to England.

You're Never Coming Back

Rain lashes the asphalt. We saunter into a bank; a machine spat out my credit card again. My wet loafers squelch on the asphalt; can the people walking by hear it?

Two wet vagabonds wandering through the warrens of the city. They have a lot of business to do and don't go home for the evening.

Matti acquires the rest of the sea charts to get us to Bergen. I cut an extra service hole in the skin of the boat with an angle grinder to check the fuel intake of tank number one tank. We check the operation of the VHF radio. We investigate why the boat's fixed NMT phone is jamming up. Everything is important and everything has to be figured out and rammed through by us, thought through under the canopy or in the rain out on the street. We have to figure things out, make calls on the portable phone, negotiate, walk, and find the energy to keep on going. With a lot of money and investment this would be different. A maintenance crew would follow us over land and be ready to take care of everything, but we're operating alone, depending on our own solutions and paying ourselves—this is what keeps me walking and thinking. I got the boat and the equipment from other people; the rest is just us and we have to figure it out ourselves, but we are resourceful and we're learning. "Figure it out" has become a part of our basic vocabulary on the *FinnFaster*.

Our Finnish flag attracts a Finnish drifter to our boat. Matti has seen this connoisseur of the art of living out in the world once before. He has been living alone in an iron boat for years, he's driven a motorcycle in the Sahara, and now he's riding a moped along the shores of Gothenburg on the welfare office's dime.

We chat about our trip with him as we bustle about, and the fellow nods. I boast about the reliability of the Evinrude and the guy spits a long loogie into the water:

"It don't make a shit bit of difference what engine you've got. You're never coming back!"

I want to get a picture with this guy, because one of us is wrong! The moped revs and merges smoothly into the flow of traffic.

He isn't the first skeptic; he was just unceremonious. Others have concealed their disbelief, smiling uncomfortable smiles, making bewildered jokes, or just silently shaking their heads. We continue our work and manage to make time for some late hamburgers.

That night the rain beats down on the canopy and inside it is wet, but we fall asleep contentedly in the warmth of our sleeping bags. We've gotten everything necessary taken care of and are ready to set course strait for Kristiansand, southern Norway, just skimming the tip of Skagen in Denmark. The weather forecast is promising good driving weather; approaching the North Sea is exciting and creates tension, but not fear.

My thoughts fly home again—our divorce will be final sometime soon. It feels distant, even though I remember every nook and cranny of our beloved home and the long life we shared together. But I don't know anything about my future. Now I'm just traveling.

The Pilot of Chicago

The high, blue-tinged silhouette of the mountains is watching us from the right. It has many secrets and has seen many dramas unfold before it on the borders of the surging North Sea. Many natural forces meet here; this sea is alive in countless different ways, and it is rarely easy.

Colorful Kristiansand is already far behind us, and we are approaching Mandal, after which the coast begins to curve northward and the open winds of the ocean extend all the way to the wall of mountains. The sun is glistening from the left, and our spray breaks far behind us; we are pushing forward powerfully in our turquoise blue world, alive and alert.

"Back there in the strait, Matti, before the engine stopped, you were telling some story about Chicago."

"Doesn't ring a bell."

"You had left the pilot at the helm and headed for your bunk when there was a thud."

"Oh, yeah, that. We were coming out of the breakwater and damn if we didn't slam into the bank.

"I ran up in my underwear and there we were, stuck. We had to rock back and forth for damn-all long before we got free again. I got in his face about it and he explained that he had to dodge something. I asked if he always ran aground when anyone came by, and he got upset and said that this had never happened before and he went on and on about how it had never gone this way before. Eventually I got fed up listening to him and snapped that if you just keep your mouth shut, it won't have happened now either. He got kind of a funny look on his face. But then he started going on about how we were supposed to file a report, and so I yelled that yeah, we are supposed to, but only over my dead body are we going to! That ship was such a darn leaky hulk anyway . . . like we were going to file a report . . . !"

Through our stories our worlds were opening up to each other more and more. We often find ourselves pondering major life decisions—our own and others—considering motives and alternatives, joys and pressures. Our perspective is broadened by the feeling of freedom in our miniature rocking world. Our distant goal makes our pulses race, but the road leading there makes us serious. Approaching the western edge of Scandinavia creates a new trembling within me.

"We'll have to look for Heikki's second-hand nautical bookshop in Stavanger."

The boat smashes hard into a big oncoming wave, sending spray flying over us.

"Whoa! Damn, what a hit. . . . Yeah, it would be nice to drop by. . . ." They're clearly getting bigger; the wind is in the northwest and we're starting to come out into the open.

I have to concentrate on the waves; they gain strength immediately at Lindesnes. The frequency of the buffeting is increasing, and my hand is moving constantly; our steady speed begins to change to a cycle of acceleration and deceleration, the sound of the engine a rising and falling growling, my eyes focusing more and more often on the shape of the closest wave.

Far to starboard ahead of us I can barely make out a downward curving, gray wedge. I stare at it, filled with a thousand different emotions. It's the Lista Peninsula, the last barrier weakening the waves; after that we will only be able to find protection in the fjords, infrequently, and after that the *FinnFaster* will meet the waves of the rising northwesterly winds, which have been gaining speed unimpeded from the shores of Greenland.

. . . Porkkalanselkä, Kihti . . . North Sea, Atlantic . . . the position of the comma has changed . . . I've groped my way here, right here, and now I'm meeting myself, my dreams and my self-assured speeches from before. I remember the fogs of the Nordkapp, the tall waves of the Arctic Ocean four years ago . . . solemn, breathtaking in their grandeur, but drivable, if you concentrate.

Rán's Daughters

Gradually we reach to the center of the drama, the waves rolling in from the fore-port side. My eyes no longer move smoothly from compass to log to tachometer to waves. I focus on the sea, glancing at the compass now and then and knowing our speed has dropped without looking at the log. My concentration makes me squeeze the wheel; my actions are mechanical as we bang our way forward.

They roll in from the fore-port side, slow in their inexorable speed, but powerful ridges nonetheless. Their slopes are alive, smaller waves surging within them. They rush towards us in an even rhythm, slowly, like the ringing of a church bell. Their slopes lift us higher and higher . . . we just keep rising. From up high we can see them all; the sea is full of them. We plunge down, down . . . focusing every second, ready for the next rise.

They aren't just this one and that one and the next one; they're everywhere and they're growing in strength.

Our flags flap above us and the wind, attacking at more than twenty meters per second, makes the ropes wail. We have to yell to each other. We fall into the gray valleys calmly; I no longer accelerate on the tops like in the Baltic—I hold back, hold back, concentrate. As we descend, the next mountain approaches, high, its peak visible only by bending my neck back. Below, all there is around us are the gray walls, between them the sky, somewhere up above.

As we start uphill we need power; we have to beat the impact of the approaching breaking crest, we have to get through it before descending into the following valley.

I push through the peaks; before the crest I turn slightly to the left—I want to meet the shock with the bow straight. The ocean explodes over us; the windshields are covered in foam. I correct our course right away. A couple of times the buffeting is too hard and we jump nastily downhill. Flying through the air makes my stomach roll; the engine bellows as the prop loses the water; we slam down from above, aft-first, on the receding slope and continue, only looking forward.

As each crest strikes the windshields, I'm also pushing through something in my soul. I laugh silently, mercilessly . . . come on out here . . . with your nice, neat lives, you back-stabbers, you gossip-mongers . . . come here and sit back on the aft bench . . . hang on, don't scream in terror . . . see those? . . . those beautiful mountains? They're just waves. . . . Matti and I are taking them on. You just go on whispering your bullshit in the office hallway behind other people's backs . . . you would faint here . . . hey, Matti, how about a round of "Old Mustachioed Boys"? . . . don't throw up in the boat; this is the North Sea—didn't you hear? Oh, you want to get back to your office and your conniving . . . that is where you belong . . . back to your musty holes . . . here you would just be dangerous ballast . . . you'd just be tripping us up . . . the sea is true . . . your tricks are a joke; I'm free, and we're enjoying our little drive . . . just two thousand kilometers left . . . for us . . . hahaha . . .

The coast looms off to starboard, six miles away. We don't want to hug it too closely; the cross-swell reflecting back from the rock wall is dangerous.

The foam of the wave crests breaking towards us glints in the sun; we are surrounded by great forces. Our own power is bolted to the back board and in our experience and attitude; even six-meter hills aren't a surprise—we take them wave by wave.

We slide down the down-hills almost idling, the sound of the engine almost inaudible; we tiptoe along carefully, when the next hill arrives we struggle up, the growling torque lifting us up and up and as the foam of the breaking wave strikes us I apply the necessary opposing force from the back board. We are making progress along our course; we don't see any other travelers.

After the crest has hit and the wind has swept the foam from the windshields, the North Sea is laid out before us in all its strength. The waves are multi-layered, stepped, rolling in from ahead to port, growing longer and taller. The *FinnFaster* pushes up to the crests slower and slower, higher and higher. Off to the horizon the sea surges with foaming crests, the bright afternoon sun glittering and the few clouds speeding quickly by.

We are here, really here, focused, but already accustomed to it. I know how out here you coast down and how you accelerate up and how you meet a striking crest. Up on top I want to look out around us, but at the same time I have to study every hollow of the approaching slope. Matti reports our course, always with absolute certainty; the compass swings wildly, but I'm just reading the average course like always before.

We only talk every now and then, the conversation cutting off up high as the water breaks over and around us. I glance at Matti. I see the familiar face, skin, beard, and beret, wet from the splashes that force their way through the gaps in the canopy as they strike. I see that he's calm and rejoice in it. I couldn't be relaxed as I drive if I felt fear at my side. I consider his serenity for a moment. His compass readings come calmly and suitably infrequently, not as continuous exhortations. He doesn't correct the readings he gives me; they hold. I can trust them, which makes it good to drive.

We climb another high slope. The crest is approaching; I concentrate, concentrate, it's almost here . . . I think about the valve. It stopped our progress after Denmark. Hans patched up the problem, and OMC's men in Gothenburg fixed it. We didn't get a new one—that was ordered to be waiting in Bergen. They said that this would last for at least that far. I push out more revolutions; the noise is tremendous. Will the valve hold? If there were ever a time it couldn't fail, it's right now! I trust them; it will hold. I trust, like I trust Matti and he trusts me; these things can't be done without trust. We break through the violent crest; the frame is strong, the Evinrude sucks its food through the valve, and the gas exploding in the six cylinders hammer the pistons rhythmically. Every turn of the propeller pushes us half a meter forward. We are moving steadily, but even more slowly.

"Hey, Matti! There's a party in Kökar today; isn't it Saturday?"

"Yeah . . . I think it's Saturday anyway . . . what party?"

"A wedding. Lasse's daughter is getting married. Anita, from Sweden. Grab the radio telephone. Let's congratulate them."

"He he, let's do it. Give me the number, and I'll place the call."

Matti places the call with his own handset. I have mine next to the throttle; I'm satisfied with this solution—we can each operate the phone depending on the situation.

"Pick up your handset; the call is ready."

I pick up the handset in my throttle hand, lowering the RPM's a little, and right after the crest yell, "Is this the Erikssons?" The fury of the wind and the howling of the flag ropes prevent me from really hearing anything. A female voice asks something. I want to get the congratulations across immediately, and all I can think of is a quick snippet of song: "Happy birthday, Anita . . ." I yell it through the thundering of the sea and hear as the phone is slammed down. That's that. I quickly put the phone in its place and concentrate on the following crest.

Our average speed is worrying me. I glance at the clock again and then the log: same thing, and it's only getting worse. It has been reading under ten knots for a while now, and on the up-hills only five or six, and there isn't any relief in sight. We are driving far too slowly; crawling along like this, our fuel could run out prematurely.

"What is the nearest protected harbor?" I end up having to yell.

"Wait . . . it's . . . Flekkefjord . . . even with us, behind the fjord on the right."

"Let's go there to wait this out . . . there's no point in grinding away here when we aren't making any miles."

"OK . . . I'm good with whatever . . ."

"Wait a sec . . . let's concentrate on the turn, so it doesn't throw us . . . watch out; I'm going to turn there on the bottom quickly and then come fast back up this same slope . . . to starboard . . . are you with me?"

"Let 'er rip."

"Here we go!"

The Lista Peninsula is foaming to starboard, the spumes rolling violently high up the slope. We rock violently in the free wind toward the mouth of the fjord. Lista . . . to starboard . . . have to drive hard, as diagonally as I can . . . hard, so the waves don't dash us against the Lista's wall . . . the valve is holding out . . . hard going now . . . I concentrate . . .

The crests coming from the port side crash against the side hull. The *FinnFaster* shudders like she's been hit with a giant sledgehammer; buckets of heavy salt water crash in through the gaps in the tarpaulin, and the foaming water streams down our necks to the drain holes in the transom.

"The daughters have beautiful shapes, but they're angry today . . ."

"What do you mean?"

"Rán's daughters, the waves . . . from the sagas . . ."

"Oh . . . yeah . . . 'trying the men, inviting them to their embrace . . .'"

It's nice to float in the fjord for a while. I find a dry cigarette. The reflection of the surrounding mountains darkens the calm water, hundreds of meters deep. We rest; the extreme concentration is past, and it's good to rest for a moment. It's Saturday evening, and in Finland people are resting on the verandas of their beach saunas. It's good there, but I feel like right now I'm in the right place in the shelter of this distant Norwegian fjord. The North Sea did not thwart the *FinnFaster*'s progress; we came by our own choice to save fuel.

The base of the beautiful, narrowing fjord continues as a canal into the heart of the small city. A strong northwester pounds the flags of a warehouse. We choose the sunny side of the docks. I take off my survival suit and look at Matti. His red-suited figure is making a round of the beach, collecting a pile of large rocks. He's taking care of the maps first; he is a real open-boat navigator. We spread the charts out on the dock in the sun to dry; the rocks are needed as weights. They don't do this on ships. I whistle. Matti's order of operation was an important signal to me again; Höfn moved one notch closer.

"Listen, about that congratulations call . . ."
"What about it?"
"I just realized that the wedding isn't for another week . . . And besides, wedding is 'bröllop' in Swedish."

Matti slaps his wet beret on his thigh. Water splatters out and he smiles, his face in a thousand smirks:

"It's always something at sea . . . some little thing . . . tomorrow we tank up and keep moving, right? I think it's starting to calm down. Have to check with the weather bureau . . ."

Pantyhose

We amble uphill along the narrow street, in the direction we were instructed. The Sunday afternoon light illuminates the colorful walls above. It's chilly in the shade; the north wind is still coming but is subsiding. Little boys race around on their bikes like anywhere else; we are just enjoying the refreshing surroundings. We don't stop at the shop windows.

I open the tidy service station door, and we enter, making our greetings. A woman replies cheerily; she looks attractive in her white, ironed blouse. I notice her funny freckles and joy-filled smile.

"Are there pantyhose on you?"

"??? How so, what . . . ?" Her cheeks flush.

"Pantyhose, the kind that women wear." She grows even redder; I start to get embarrassed too. Suddenly I realize! I'm using the words as if I were speaking Finnish.

"For sale, I mean. Do you sell them? We need to buy a pair."

Everyone solves the puzzle at once, and we burst into sweet laughter. She continues, a little confused:

"Yes, I have them, but . . . what size . . . how many denier, and the color . . . ?"

"It doesn't matter; just any."

"How can it not matter? They're all so different . . . Is she . . . ?"

Matti turns away, shaking with laughter, and I'm starting to catch it again too. The woman is so sweet and helpful, and here we lunkheads didn't have the sense to explain up front.

"Oh, hey. It doesn't matter; they're going to be a filter."

"A filter?"

"Yeah, a filter. We're filling up our tank and the filter plate in our big funnel has gone missing. We're going to make a filter out of pantyhose; it'll work really well."

Matti goes off to build the filter, and I stay to arrange the fill-up. The fuel has to be carried four hundred meters in canisters; they have a cart.

"How much do you need?"

"About three hundred liters."

"So much?"

I tell her about our trip; we chat for a long time. She listens carefully and asks a lot of questions. Whenever we look at each other, smiles bubbles up on both of our faces; it feels good. The sea is rolling along out there, and Höfn is waiting somewhere, but right now I want to stop and smile.

"That would be so hard with the cart; take my car," she says and offers me the keys.

I drive our gas to the dock in the clean, new minibus; it takes more than a few trips. Her willingness to help moves me; she doesn't even care that there isn't any way to avoid some of the fuel splashing on the carpet in the back.

"Don't worry about it. It hasn't been easy for you to get here, and you have a lot ahead of you. Just go ahead and take it."

I return the car after a couple of hours and pay, but I can't stand to leave. She's so refreshing, captivating. I wander around the shelves, as if I'm looking for something. Is she wondering what I'm up to? I glance, and she glances back. Why am I wandering around here? We look at each other again, smiling at the same time, and she blushes, but her eyes don't move away; she smiles straight at me, boldly.

I buy some cigarettes and sundries. I pay at the counter, lingering.

"Do you have . . . to leave . . . tonight?"

She whispers it quietly, looking me in the eyes. Our hands brush each other and remain close. I see all of her: the short, blond hair, the freckles, the exuberant, open smile, the straight posture and buxom bosom. Under her white blouse her nipples are hard, shamelessly ebullient.

A thousand ideas race through my mind. A woman . . . radiant; she asked me . . . the sea before me and behind me . . . so much . . . she's nice to me, she likes me . . . Lord, how I want her . . . I am divorced, probably . . . a smile . . . for me . . . Höfn is a long way away; we have to drive whenever we can . . . Muru, Muru, you're so far . . . away . . . waiting, encouraging . . . I know . . . Muru . . .

"Yes, I have to leave today. There's a storm chasing us, so we have to get away in time . . ."

"I understand. What's your name?" She asks seriously, quietly.

"Pekka."

"I like you, Pekka. I'll be thinking of you. Go now."

Hold On

The mouth of the fjord is an immense celebratory gate into the unknown. The high, shadowed wall on the right side rises up into the sky, the bright sky reddening above it. The setting sun tints the mountain slope to the left a soft, dark orange. Beyond them and between them lies the swell of the North Sea, powerful but no longer frothing.

We push our way forward toward the opening and then stop. The pressure of yesterday is in our subconscious; we concentrate, gathering our battle lines here in the calm of shelter. We chat quietly, at long intervals, and go over our driving plan that will take us to Bergen. This will be a long leg, almost two hundred miles, but we are going to try to slip through before the strong northwesterly wind manages to gather strength again.

"Let's keep the same distance from the shore as yesterday. About six miles. It's still blowing and farther out its calmer driving."

Matti is bent over his map, counting out loud with a set of calipers in his hand and wearing his reading glasses:

". . . the first waypoint there, that makes . . . ok . . . turn, about . . . and then . . . next . . . about there . . . then to Stavanger . . . there we could check . . . Yeah, ok. Six miles is good for me too; don't have every point messing things up . . . did you get a hold of Harry?"

"Yeah, finally. Promised to meet us at the fish market . . . asked us to try to contact him after Haugesund . . . and asked the news from Kökar. He's coming to the cabin in July . . . Harry is a good guy; you'll see . . ."

Everything is ready, but we still don't start off. We lean on our headrests, looking at the opening ahead of us. We're thinking the same thing, measuring this long stage of our journey in our minds, anticipating what is to come and collecting ourselves. The engine is idling nice and steady. This is a very communal moment.

"Let's go . . ."

"Hit it!"

The *FinnFaster* accelerates in the dark, smooth fjord into a strong hydroplane, only the aft part of the hull cutting the water. She speeds toward the opening for five minutes and then slows and meets the powerfully surging cross-swell at the mouth of the opening. Water slams against the sides, shooting high into the air. Then she begins to settle into the waves and directs her bobbing path to the right, curving gently. The wake left in the fjord has smoothed to nothing.

"Take five more degrees to port."

"Five to port."

I'm following the compass . . . no-ow . . . went over, turn back . . . there. The compass swings restlessly; the sea is bumpy, the troughs growing. To Bergen . . . a long road . . . have to get driving . . . there's no way . . .

"It's still so strong, so alive."
"Feels like it. Is it hard to hold course?"
"No, but I can't get any speed . . ."
"OK . . . you know best."

The hours crawl by; the sun has fallen below the mountains to the right and the night grows dark. The sea is difficult—two or three meter, steep bumps, hitting from fore-port, sharply. I hold the boat on course and try to avoid hard hits; we don't really talk. Now and then I roll my shoulders and check the log; our speed is under ten knots, only getting up to eleven momentarily. The bumps grow stronger, and I start to get nervous. We aren't making progress, and I'm getting tired. Matti is quiet too.

I illuminate my wristwatch with a flashlight—I taped it to the handrail next to the instrument panel. Dammit! This is getting us nowhere! Time is passing, a new storm front is approaching, and we're burning too much fuel without covering enough miles. Hell no!

I stop. Matti looks at me inquiringly.

"What now?"

"Wait a minute. I'm thinking."

I close my eyes, listening to the rhythm of the waves with my body, trying to work myself into them; I know how to do this. It won't work like this; what's wrong here is me. In a flash I realize, I remember, but above all I perk up. I'm sitting here behind the wheel so tired that I'm drooping and trundling along like we're taking a Sunday sleigh ride to church. I remember how once . . . many times . . . that's what I always do with my own gear in this swell rhythm, if I'm in a hurry.

"Matti, hold on tight. I'm going to try to get this girl moving."

"OK. Let 'er rip; I'll be fine . . ."

I accelerate normally, trimming the bow up, and the impacts grow harder.

I'm still accelerating, the bumps pounding violently against the hull. Still a little more. I look at the waves intently, giving it gas . . . not too much . . . now . . . still more quickly, the bow down a little and my throttle hand light . . .

The *FinnFaster* rises, roaring, the waves hitting her at an accelerating pace, but the blows are soft against the bow. The boat speeds horizontally across the peaks of the waves.

I glance at the log. 11.7 . . . 12.4 . . . more . . . 13.2 . . . 14.7 . . . briefly over 15 . . . hell, yeah, it works here too. Ha ha! Now off to Bergen instead of moldering here.

Our pace is furious given the conditions, but I have the boat under control. I have to take the jumps carefully, the downhills lightly on the stern, so we don't hit the next slope with the bow; the gas, steering, and trim all have to be adjusted with millimeter precision, to the second. The relationship between the swell, the boat, and our power is ideal for driving on the peaks, bucking with this rapid, cyclical roaring of the engine up here as the bow shoots foam far to each side. I rejoice, remembering one time returning from Harmaja years ago with my first boat; I learned it then, when I tried . . .

"Dammit, Pekka, I didn't know this thing could fly too. Damn!"

"Well, almost at least . . . like that, good . . . take-off speed . . . a little more . . . now, there . . . try again . . ."

My fatigue is all gone. I'm just enjoying our dash through the darkness of the night. Our running lights turn the spray on the starboard side green and the port side red. A mysterious, roaring creature speeding across the tops of the angry waves. Despite our pneumatic seats, we have to take some of the shock with our bodies, but we are finally making progress. After every slowing I jam the throttle forward so we don't fall down into the chop. Our spirits are high.

"Old mustachioed boys . . .

" . . . man and seal both they be . . . !"

Do It Again

Our roaring dash continues for hours. Concentration banishes fatigue, adrenaline boiling through our veins. Our high speed and the crosswind set our flags sputtering; the water roils beneath us. I think of our aluminum boat as the body of a large bass violin being played violently. I listen to the sounds, enthralled, throbbing with the huge sprays of water, accelerating at the very limits, ignorant of my age.

Those sounds, those wonderful sounds . . .

"THE SOUNDS, MATTI . . . LET'S MAKE A REPORT!"

"You mean now?"

"Now, now! Right now. Order a radiophone call!"

"Hehe . . . why not . . . they can hear a little something . . . tell me what number."

I'm excited. YLE will get a report straight from the North Sea at night, for the nighttime broadcast. This is what they wanted—direct and authentic. This is certainly that: A small Finnish boat, two men, the night, the waves, the hellish roaring, yelling to talk, the big engine screaming and the sea pounding. This is our journey in sounds. I think of people sitting around a grill in Sipoo drinking in the wee hours. I think of the small farmer on his way back from courting in the middle of nowhere central

Finland. I think of everyone up late at home . . . this will wake them up!

"The call is ready; can you handle it while driving?"

"Yeah, I can manage. I'll talk as much as I can."

The call connects to the studio in Pasila. A young reporter answers.

I introduce myself and quickly state my business. I juggle the wheel, the throttle, and the handset, following the waves closely. I'm flying high right now—I'm about to give a powerful report from the heart of the action.

The young man in the studio is confused and says quickly, "Call back again in five minutes; we're in the middle of a disc."

I am struck dumb and gasp for breath for a moment. In the middle of a disc; call again . . . ! The boat lurches. I correct with my left hand and soften the blow with my body.

"Smash the goddamn disc against the wall, boy! You were just about to get the report of your life." I slam the receiver down.

Matti looks at me inquiringly. A bucket-worth of spray slams on the windshield.

"It looks like the radio is all kiddie disco now. They were in the middle of a CD—no time for a report just now, if you can believe it."

I am agitated, furious. What is life about for them? Are we supposed to live real life between hit records? What's so sacred about discs? Do adults really listen to that too, the charts and rankings? Is that enough, monotonous pap that blots out thought? We went to a lot of trouble—we even took a risk with the handset, just wanting to share a very real moment. Call back? After some young broad has first gotten her chance to whine her vacuous ditty . . . then call back from the North Sea . . . and this is public radio . . . the taxpayers' money . . . goddammit!

We can't get another connection and the driving conditions are getting worse. I have to concentrate completely, but our speed is holding. The situation with the phone keeps drawing my thoughts back to life at home. I keep asking myself who makes those wheels spin.

The light of dawn reflects palely on the tops of the mountains. The distant, black silhouette is sharply outlined against the sky; Stavanger is fifty miles away. We've checked our location and wolfed down some dry rations, added oil to the reservoir and switched over to the full tank. The difficult conditions have remained unchanged, but we have gotten used to it. Nothing bothers us and we feel like we have the upper hand.

My driving is routine, although at the same time delicate, quick adjustments by hand, softening hits with my body and giving quick commands to Matti:

"Watch out, jump ahead . . . no-ow . . . moving on now . . ."

The rhythm of the waves is in my body; once again I am part of my boat. I drink in our progress and get more and more excited as we approach Bergen.

Matti belts out, "Wreetched waandering seeaman..."

I'm taking a crest normally, glancing at the log at the same time, but this crest is bigger than the others. I don't manage to give it enough gas, and it sweeps us along with it. We speed along on our right side in the froth of the crest for a few hundred meters, the engine braying, horizontal in the air. Green foam churns below me, and Matti is sitting to the left, holding on with two hands. My thumb presses the trim switch; I hear a buzzing and squeeze ... not yet ... down, down engine ... soon ... we'll fly along with the crest ... so far ...

Finally, finally I feel the prop meet the water; I jam the throttle forward and throw the wheel to the left. I ram down through the foam; I calm the boat and then we are already taking the next wave, continuing on as before.

After a moment, from the midst of the silence I hear Matti's rough voice.

"Shit, Pekka. That was good ... do it again!"

We explode into laughter. I declare that there will not be a repeat performance; this was a special, one-time engagement. Matti explains, his face serious, that, "You did it so softly; it was good ... like calmly correcting a car sliding on ice ..."

The opening of the sound to Haugesund begins to come

into view, and the orange glow atop the mountains is blazing at full strength. The good feeling sends my thoughts flying home again; wake up feeling well, all of you I love . . . we'll make it in by the afternoon.

The change in course returns us to the normal way of driving. The mouth of the fjord is an hour ahead of us; calm waters await us there. I'm still thinking about Matti's comment after we flew on our side like that. The nerve of most experienced sailors would have failed. I can go far with a guy like this—no need to worry about the navigator being afraid . . . do it again . . . damn!

The Scent of Summer and Winter

Leaning on the headrest is divine. The crystal fjord stretches out like a mirror before us, the blue mountains bounding it a wedge tapering into the distance. The sun rising from low on the horizon warms our faces; our pores breathe in the scintillating light. The engine, shut off, clicks as it cools, but in our ears we still hear its ten-hour stretch shouting and bellowing. Our bodies are still trying to soften the blows, and the hull is still rumbling like the bass in a violent symphony. And that is precisely why we are princes. If we had been brought here to this glistening paradise on a tourist ship we wouldn't be experiencing anything—we would be outsiders.

It is fifteen minutes to seven—no rush anymore. We can weave through the smooth fjord luxuriously, seeing the villages wake up, meeting ships and boats, greeting people. We are curious and excited; everyone is our friend. This is so easy . . .

I startle—I smell a familiar scent! The luscious aroma of a summer morning combined with the crisp cold. I have experienced this twice before. Five years ago when I was on the run from the pressures of my life in Zug, Switzerland, I stood alone on a dock on the small lake with the mountains rising around me and looked at the glamorous boats that had been imprisoned on that pond. It smelled of summer and winter together. Four years ago the same smell wafted over the island of Skjervöy in northern Norway. Elina and I enjoyed it together in naked languor on the edge of a glassy Atlantic in our gently rocking boat. The snow caps around us rose a kilometer into the sky. Nordkapp was behind us then . . .

"Do you want some sausage?"

"Yeah, whack me off a good chunk. There should still be some chocolate too; we should be able to get by another six hours on that."

We eat leisurely; the terns get some of stores our sausage as well. Their cries are homey—they have the same language here as at the Market Square in Helsinki.

"Magic time again?"

"Yeah, but wait a minute and let me look at this some more. This looks like an interesting slalom . . ."

"That it is, and there's no way you could ever do it in that ship of yours. Remember, Matti, to yell stop immediately if you need it; I'm relying on you completely as I drive, even if we are in here and I can get her up to twenty-seven now. You order the stops and the lines we take, and I play the organ."

"Hit it!"

A few hours later we get a whopping idea: Let's go on an outing, a picnic! We drive up next to a flat islet, tie off the ropes, and set up on the glaciated bedrock. Our bag of provisions and a couple of bottles of beer make the mood festive. We wouldn't trade this with anyone. We lounge around for a couple of hours in the warmth of the sun and laugh about how comical the situation is: Two old men in an open motorboat driving one thousand five hundred miles around Scandinavia to fry up some sausages on a tiny rocky island in a fjord in western Norway. Are we walruses or wild boars? What does it matter? Plenty of people think we're insane. What the hell does that matter either? We're just brazenly hoarding the best pieces from a table we set ourselves. Let everyone else do what they want to do—or know how to do! We think today is Monday, but we aren't sure and don't feel like checking.

The engine is already running; Matti is coiling the bow line and we are free of our little islet. The radio crackles: "Attention all vessels . . . attention all vessels: Storm warning . . . storm warning . . ." We glance at each other.

"No problem for us."
"No problem. Anymore."

My Bergen

The din of the high, arcing bridge over us says that it is rush hour. We are crawling along at an idle toward the center of town. The rapid surges of fog and driving rain rob us of visibility, but through the clouds we catch glimpses of the colorful houses on the inhabited mountainsides and of the valley, which has supported active habitation since ancient times.

Bergen, the pearl Viking tales, lying between the mountains, my Bergen! I am here again! We head toward the Hansa dock as the exhilarating memories roll over me . . .

. . . an exciting car adventure alone with my sons . . . Elina hugging us as we left . . . we were still a family . . . Perttu had just finished his college entrance exams; he got to drive a lot. On the downhills, 13-year-old Arttu yelled, "If only I had a skateboard." My stomach was killing me . . . a bivouac in an intake center for Kurdish refugees . . . dead tired boys, tenacious and loyal . . .

. . . we breathe in the history of the Hansa dock; we find exciting things; we shout to each other in glee and are all the same age . . .

. . . Harry picks us up at the cafe we had agreed on . . . the friendship, which began on Kökar in the shelter of Nestor's hut on the shore, like minds met . . . the building site; he offers it to me . . . I begin to plan the cabin . . . Harry and Åse, in Bergen . . .

. . . The smell of history and of the Atlantic is present, the Vikings, the Hansa ships, the whale hunters . . . the tumult of the smoky pubs . . .

. . . finally a cluster of pearls of light, after twenty-five lonely hours of driving over the wintry mountain range, I call Vivi . . . I'll be at Harry's place soon, at my destination . . .

. . . the surprised trickster editor-in-chief when I show up at his dark hole to collect my money, when I jerk the door open . . . I was warned . . . gangsters . . . still more surprised when I close the door behind me . . . I'm afraid, but he is more afraid . . .

. . . the smile of a dark beauty during an erotic art show on the same trip . . .

. . . the stern eye of the old guard at the seafaring museum . . . an old skipper . . . I ask about the swells here . . . the spark is already alive . . .

. . . the smirk of the commodore of the yacht club . . . I wanted to meet someone who had driven to the Shetland Islands in a motorboat . . . he snorted as if he was talking to someone who didn't know anything: "There is no such person . . ."

My Bergen is close; I can smell it. There is the Hansa dock. The fish market! I am coming by boat, far away from Helsinki!

The dock is a couple of hundred meters away; a cloudburst opens up for a moment. On the dock are a few figures under umbrellas.

"Goddamn it!"

Matti points at the windshield; the drops have frozen. Tomorrow the school children will be singing "*Jo joutui armas aika...*"[1] at their graduation ceremonies . . .

A mustachioed, straight-backed man hands his umbrella to a youngster, hurries to the edge of the dock, and nimbly catches our rope. The grin is familiar—maybe there is a little more gray around the temples.

"Howdy, Pekka. How did it go?"

I look at Harry for a long time. I have a friend here too.

"Vivi says hi."

---

1 A hymn traditionally sung in schools at the beginning of summer vacation. Also known as "Suvivirsi" or "Den blomstertid nu kommer" in Swedish. The traditional lyrics extol the beauty of God's creations at the coming of summer. In secular settings sometimes only the first verse is sung or alternate lyrics are used.

A Pitstop and Edvard Grieg

Åse is humming in the kitchen. The mild smell of fish soup teases our hungry senses, and dishes clink as she sets the table. Harry rolls a cigarette and then fills our glasses. The living room is like before, bright and clean. The framed photograph I sent for Christmas has found its place. It shows a fishing boat gliding through a maze of islands after sunset, with Nestor in his cap at the helm, a dozen years younger.

"Vivi is doing well, I assume? Has she planted her potatoes yet? And Hemming, he's keeping Karlby in herring in his free time and going flying now and then? I heard about how he fixed up his old dad with a vacation day from the rest home." He laughs.

We pass a congenial afternoon in Harry's home in Bergen trading news. We stay for a long time—it's nice to be in a real home for a change. So many good people and things connect us, and we know they have been looking forward to our visit.

We return to the boat to sleep. We don't want to get used to sheets—it would be too easy and too dangerous.

"Great news. Your engine is in tip-top shape: All of the gaps and tightnesses are as they should be, the transmission oil is crystal clear, and there is no oxidation anywhere. The plugs were worn out, so we switched them.

We did replace the three-way valve, and here is a diagram of how the new one is set up. How many running hours are you up to now?"

I look at the installer thankfully. He smiles a lot and clearly knows what he's doing. On the back of his blue overalls it reads "Björdal & Madsen," the name of the shipyard, which has an authorized OMC shop. A wave of relief rolls over me, simultaneously a wave of trust and faith . . . tip-top shape, excellent!

"Almost exactly one hundred hours since we left . . . and most of that has been pushing pretty hard."

"This was good time for an overhaul then. It's a wonder she was still running so nicely with those plugs. Here's a backup set and you can get filters from the store . . . there, from the shelf on the left."

I think of Hasse Johansson with appreciation. OMC Finland has done a great job setting up their support; they gave me a list of all of their service locations with contact information and people all along our route. The engine will get the proper care. We just have to survive to the next station.

"Piri, your fax came."

Matti and I begin reading the story the young reporter intern has written. I met him a week before leaving and stipulated that I get to review the text. He called and I gave him the shipyard fax number. Apparently there was a huge rush to get it back—of course.

"Heck no, where is he getting this? This is like from another planet. Doesn't he understand anything? "Risking death?" I said a hundred times that this is a controlled adventure, that we aren't thrill seekers, that this is a professional gig, that we aren't just some kids out for a lark . . . The boat has "automatic pumps?" No no no. Trying to "sail" to Iceland . . . sail? Matti, what are we going to do about this? This is complete bull. Almost everything has to be rewritten. He's disgracing us."

"Don't take it so literally."

"Literally? I'm not letting anyone write shit about this. Apparently we don't even know why we're doing this. Is that true? Is it?"

"What, is that in there too? Of course that isn't true!"

I make a pile of corrections. The fax paper is like my German exam after Mrs. Anja Virta graded it—I got a one minus on it because they didn't give anything lower than that. Then I call the reporter. The connection is poor, but that isn't the only thing that makes me raise my voice.

". . . and not a word about automatic pumps. Where did you dream that up?"

". . . but the bilge . . ."

"This is a self-draining boat, just like I've already said over and over. Will you believe me already!?"

"Isn't it the same thing? This is just nitpicking."

"No, it isn't. It isn't even close to being the same thing. You just write self-draining. I don't have time to teach you Boating 101. Just believe what I'm saying: S E L F

D R . . . If you'd just listened when we were talking. That was why I made time for you. You can't just make things up about the topics you're writing articles about. At least not when it comes to this trip!"

Matti climbs closer to the doorway of the crypt. Bushy, wet mats of ferns brush against his legs. A vertical concrete slab poured directly into the dark wall of the cliff closes the crypt above him. The lush, wet forest and undergrowth at the base of the cliff create a deep quiet. Birds twitter high in the tops of the trees.

We've suddenly been catapulted into another universe. The director of the shipyard wanted to show us the nearby tomb of Edvard Grieg and his wife, a shared resting place mined into the side of a mountain. It is truly impressive.

The red beret stands motionless amidst the green. A serious face gazes intently at the mountain crypt, bushy eyebrows not even twitching. Matti is experiencing something important.

". . . *Peer Gynt* . . . 'Solveig's Song' . . . they've always moved me . . . now I'm suddenly here, right next to him . . . Pekka and I have come from so far . . . the master composer . . . to pay our respects . . . The record I listen to at home in the winter . . . 'The Dance of the Mountain King' . . . right here . . . he rests within his own mountain . . . this amazing voyage of ours . . . a solemn moment like this in the middle of it all . . . we're going on . . . the interviews, the pressure . . . I can take it . . ."

I stay silent, motionless. I understand the moment Matti is having and respect it. I'm happy for him. I carefully wind my camera.

The hour we had agreed upon has almost passed. We walk in silence to the waiting car, a little bit stronger.

A blue pickup truck pulls up to the dock at speed, gravel spraying when it brakes and dust rising from the wheels. Harry jumps out and drops the tailgate. He is full of energy.

We position the two-by-fours on the deck of the bow, check the measurements, drill, bolt, and nail. According to our calculations, we need more tank space for the last leg of three hundred miles, so we're getting things set up here in the big city. Here we can get the right parts, and here we have Harry.

He has thought of everything—this is an easy job for him. He drew for a moment last night with his sure hand, calculating out loud, taking measurements, and speeding over to our boat at the arranged time after his workday.

The support stand is ready in no time. We position two 200-liter barrels lying down side-by-side in the bow. Both have valves with sturdy handles. After the boat's front tank empties, we'll drain a barrel down into the drive tank using gravity, and then, later, the other. We're able to carry 400 liters more now, 1200 all together, which the *FinnFaster* and our Evinrude can handle just fine. The boat's basic handling characteristics will still be the same, even if it is a little tanker now. It would be pointless setting out in a barge.

I watch Harry at work. Everything works the first time and fits in its place. He is quick, sure, and precise in his movements. A childhood spent in the austerity of the outer archipelago has taught him that you have to get along on your own out there or "you die." The standard phrase I started hearing more than a decade ago, "*inget problem*," is repeated frequently.

"Hey guys, are you ready?"

Tall, dark-skinned Åse has come to the dock. Her pronounced Bergen accent with its strong R's sounds rakish, and she often laughs with her whole body. I prefer to call her "Åse the Viking," which she likes.

Åse sets out the coffee on the deck barrels. The black coffee from the Thermos and the princely pile of open-faced fresh salmon sandwiches in the mild early summer evening give the occasion a festive feel. The *FinnFaster* is now fitted for ocean going.

The Frenchman is clearly proud of his boat. He arrived at the fish market dock the day before yesterday. The 15-meter open sea sailboat has all of the niceties. He invited us for a visit after hearing about Matti's profession. Their wine is good, but I'm embarrassed.

The woman is reserved towards us; it might be a result of our dirty jeans and squelching shoes. Our host is wearing a white sweater, his dark hair is well-groomed, and he smells of expensive ointments. They are considering whether to hazard going up the coast to Trondheim.

"Look, I've figured out your route to Shetland. The day after tomorrow will be the best time to go; I looked here . . . there are a lot of graphs here . . ."

He points often to the weather fax on the wall of the saloon, which pushes out paper and barometric isobars whenever he presses a button. We're only half following him and getting bored. The lady doesn't really speak at all; she was expecting guests in uniform. I'm growing impatient with the elegance. I think and think—and come up with something. I glance at my watch and then stand up: "Thank you so much. I'm sorry that we have to leave. Finnish Broadcasting is waiting for our report. We have an arrangement . . ."

Matti jumps into my wake like lightning, and we're out in the fresh air. We try to stifle our laughter.

We're waiting out the rain under the canopy chatting about our departure. Matti is measuring the distances; he looks at me over his glasses and then slaps the logbook shut.

"We're leaving tomorrow, no matter what Frenchie says. That's my opinion."

"I don't want to hang around waiting for the front either; we'll just get out from under it. All we need is twenty hours. The weather is the navigator's business on this boat, and that's why you're there. I trust your evaluation."

"The front is coming faster than our neighbor thinks; he's a bit heavy on the theoretical side in this stuff. It'll be on us the day after tomorrow and will last for several days. We've gotta get out of here quick while we can."

"OK, let's go. Absolutely."

The fish market at night is beautiful. The thousand lights of the city reflect on the wet asphalt and everything repeats like a mirror on the bay. I look at my panel silently, absorbing the lustrous Bergen night. Our boat's sturdy targa rack barely extends up to the level of the dock. The antennas sway; Matti is moving around under the canopy. A taxi swerves violently toward the city center. The French have their flags flying.

We're leaving tomorrow. Tomorrow the bow railings will be pointing towards the ocean; the shore will fall behind us and we will be alone, out of radio contact. I feel devotion, not fear. I have waited so long . . . the calculations, the estimates . . . all of it will be tested from here on out. Höfn is 700 miles of ocean away, between us only two ports.

The hour is late. I want to share this moment, and my thoughts fly to Muru. I'm already dialing the number.

She answers drowsily, but her voice is calm and clear. The familiar, low timbre reverberates in my ears.

"It's Pekka. I woke you up."

"Mmm . . . honey, where's my wanderer?"

"We're leaving in the morning."

She wakes up quickly.

"West?"

"West."

"Oh, Pekka, I feel so good. I remember everything you've said about it. You've already driven so terribly far, and the journey just goes on . . . because you want it to. I'm thinking how you must be feeling. You're so close . . . hehe, you'll never guess!"

"What?"

"Tonight I wanted to put on a special nightgown."

"Oh. Which one?"

"The blue one. Of course the blue one."

"Muru, listen . . . can I lie with you tonight?"

I hear a low whisper:

"Come be with me . . . mmm . . . oh, Pekka . . . come."

Matti is snoring lightly. Rain is pattering on the canopy. I am burning a candle on the steering console and writing feverishly—my pen is flying. I am continuing our phone call. I want to tell her everything, just to be sure. I want her to have it in black and white, saved away somewhere. The letter leaves for the East in the morning, we for the West.

The Special Nightgown

The morning begins calm, cloudy but dry. There's no sign of the alley cat. It had jumped on board in the night next to me when I hung up the phone. It made its way right up to me, and I gave it some sausage. I thought of Muru and our journey, of Elina and our sons, of the whole intricate path of my life, the current twist of which I was experiencing here on this dock in Bergen in the middle of the night. I scratched the cat and it purred a little.

Matti will be arriving soon. We're leaving in half an hour. I find it relaxing sitting here on my dock pylon. The center of town is buzzing and a black Hurtigruten cruise ship is turning in the inner harbor, belching smoke from its stack. The city continues its buzzing. Our departure won't have the slightest effect on it, but for us it is a big thing.

I decide to call one more time. Now is a good moment.

"We're casting off in half an hour."

"Excellent. What is the weather like?"

"Good. Mostly tranquil, cloudy but dry. We're trying to hurry out from under these clouds. There's a new front moving in. We need about twenty hours. I just have to find the energy to drive."

"You'll do it."

"Muru, listen . . . how did you sleep . . . in your special nightgown?"

"Hmm . . . hrmph! How special can a nightgown be... if it's still on in the morning!"

"You're a sweet vixen, Muru."

". . . well, Pekka, you listen now. I've got a meeting starting here presently. You just pop over to Iceland now and then we'll see each other, right? Hugs and SMOOCHES!!"

Matti arrives full of energy. He claps his hand together and yells, "Let's get moving! Let's get over to Britain finally so I can do a little more talking too."

"Right on. Let's go say hi to the lords and ladies . . ."

The Evinrude is idling. The fenders are in the boat. Matti has untied the bow rope and is tightening the knot on the canopy bundle. I back away from the dock slowly, turn, and shift into forward. My survival suit feels homey again after three days in the city . . . ay ay ay . . . off we go . . . to listen to the song of the engine . . . and our own . . . Out there set against the open sea is the long island of Sotra. A friendly merchant is waiting for us there; he has gotten everything ready and he has a lot of fuel; we've checked.

Leaving Scandinavia

"It's getting oriented bit by bit."

Matti punches some buttons and looks at the map.

The GPS satellite navigation reacts slowly to turns. I watch its readings and check the magnetic compass—I'm trying to line up the boat on a fork. As I hold the right heading long enough, adjust, return, and keep driving, we move slowly along the baseline drawn on the map two hundred miles into the west, toward the Shetland Islands.

The shore of Norway has disappeared behind us. The cloud cover breaks now and then, and the weather is still mild near the shore. The tingling of excitement of leaving is gone. We've settled into our posts. We're back in our rhythm. As the shore recedes, the boat grows. It becomes our whole world; everything is here, everything important, framed by the clouds, the fog, and the ocean.

"How did people react to you leaving, Pekka?"

"Oh . . . in a lot of ways . . . yeah . . . this is one of those things that divides people . . . sort of sorts them . . . I've heard everything, the full spectrum. It's been interesting . . . some professionals were excited, thinking through everything we would have to do, and then of course the landlubbers just started hyperventilating. Some of the boater types were as silent as the grave—this has probably pulled the rug out from under them . . . It takes them a million euros worth of gadgets to just barely make it to Naantali and then stagger around the casino playing admiral with a glass of company whiskey in their hand . . . I imagine that is a bit of a challenge . . ."

"I had some of the same. A lot of people just clammed up . . . except for one pilot from Rauma . . . he actually shouted and wagged his finger at me . . ."

"I had a really interesting conversation on the phone a week before we left. This one lady gave me a real earful."

"About what?"

"Well, she went on about how I had to either be in extreme need of attention or god-awfully money hungry. She was really fuming and went on that there wasn't any amount of money that could get her to do something like that. It made me laugh so damn hard, and I wanted to ask who the hell would ask her to try . . . Then she was making snide remarks about how we supposedly love the ocean so much but the boat is coming back from Iceland on a ship and she'd read in the newspaper how that singer, Lasse Mårtensson is spending his whole summer in the archipelago in his sailboat . . . and this is costing me an arm and a leg . . ."

"I would imagine so . . . what a load of crap . . . not everyone understands everything, even if they do have an opinion . . . about other people's lives."

"Yeah . . . I didn't bother listening to her finish; I just put the receiver down, even though it felt rude. All that was missing was for her to wish that we would drown."

"Yep, I guess so. How many revolutions?"

"It's resting right around four and a half thousand—has been for a while. The log says 17.3. Which is just fine."

We're skimming along the gray, two-meter swell steadily, no change in course. Now and then the bow splashes deeper after a higher peak, water flying past us; we struggle up the next and the next in a tight plane, continuing through the emptiness. The growls of the engine are accompanied by the banging of the hull.

"The highway view on this GPS sure is nice. I'm driving by compass but then checking against the GPS. It's great on the long stretches. Gives me a general picture."

"Yeah, that highway is good stuff. It's easy for me to follow how things are going too."

"Talking about highways just made me think of something."

"What?"

"It's from my old job, this one trustee, a lady. One spring night in a taxi, she started throwing herself at me, really forward. She said something about having a weakness for Aries. We'd been drinking on the company's dime, supposedly as part of negotiations between the employees' union and management—corruption shit. I moved her paws off me and pretended not to notice, looking out the window."

"What then?"

I wipe the fog off the windshield and add a hundred RPM, getting us up to eighteen knots.

"Well . . . later the turkey showed up at my door. She came with her delegation and stood there shaking telling me that things were going to go back to the normal routine in the company. That the personnel manager would decide exactly like the shop stewards ordered or else it meant the highway. She was adamant."

"What did you say?"

"Well, I knew the score, and it was a tight spot. Then I said back that it was good she was talking straight, that I liked that. But she had forgotten one thing. That she couldn't pressure me, because I wasn't afraid. So get lost. It was a short romance of sorts. Doing it the trade union way. Haven't missed it much."

"Hey, look. Now you're on the highway. And it's looking good!"

"Hell, yeah . . . no kidding . . . how did that song go . . . the Ostrobothnian one . . ."

I push against the footrests and stretch my body out firmly against the backrest. We're diving through alternating fronts of fog and rain and haven't talked for a few hours. The silence doesn't bother me. If Matti wants someone to talk to, I'm right here.

The hull bangs away rhythmically and the spray to the sides rushes by. The position I need to be in to drive doesn't let me move around freely, and my muscles are numb. I try to support every square centimeter of my back on the seat. I feel like I'm going to need every last drop of my strength. I consciously save what energy I can. We've driven six hours from Bergen, and we have the same amount left to the destination we're sighting towards off the Shetland Islands, and from there a few more hours through the craggy mountains to the harbor.

I look at the surrounding grayness, considering it, and it swims inside me. The feeling starts in the pit of my stomach, spreading like blood circulating through my body, into my chest and throat, finally blasting in my brain and consciousness as trumpets of joy. I am here, right here, right now. This is true! Our whole 1600 mile wake to this point is true. We really have gotten her, one wave at a time, and we will go on. Somewhere in the distance, before us behind the fog and mist lies Lerwick, sleeping: a pub door creaks, a dog barks, and a clock strikes the midnight hour.

"Look at that gull, Matti. It's been following us for almost an hour."

"It isn't a seagull; it's a storm bird, *Fulmarus glacialis*. They're shearwaters, like a northern albatross."

"Ah . . . Looks just like a big gull."

"Look at its beak. There's a lump, a sort of tube hole. Those are normal for long bills. It uses the tube for its salt system. I've seen thousands of them. Maybe more."

"You know birds."

"I've always been interested in them. Nesting and migrations, all those systems. If I didn't sail I'd probably have been an ornithologist . . . sometimes I really would want to be . . ."

I glance at the log. We're still at a good seventeen knots. I'm doing okay keeping our speed up. I see my wet moccasins lying in the foot well. They've been squelching since Visby. The bearded doorman at the bar in Bergen laughed like Pavarotti at the sole that had soaked loose. I pick them up in my hands and toss them over my shoulder into the wake without turning around, first the left, then the right.

Matti glances over from his radar:
"Oh?"
"Yup."

We stop to take our bearings. At the same time we fill the oil reservoir, gulp down some food, and rest for a minute. The silence of the motor and the slipping of the boat into the gentle rocking of the surge as we drift free gives us a moment's rest.

I do some exercises, stretch, rinse my face, and try to make the most of these ten minutes. I have to.

The darkness is closing in. It's almost night. We are ready, but I don't start the motor. I just look around, and Matti doesn't rush me. He knows that I am the one driving the boat to Shetland—that I need this moment—so he is patient.

The sunset colors the north with a silver cast. Between two darknesses is a glistening strip with a hint of violet at its edges. The horizon has disappeared. We rise and fall. I feel a joyful wave of timelessness; I am thankful and humble. I would never trade my mountaintop for the security of the flat country, even though I know that the price will be an agonizingly deep valley somewhere ahead.

The boat is jostling around violently, and the rhythm of the hits is confused. We are approaching the shore and the currents of the millions of cubic meters of ocean water around the islands, the increasingly shallow bottom, and the rapidly rising wind are intermingling in an angry dance through which our small boat is struggling forward more and more slowly. The humidity has fogged the windshield completely, but that doesn't matter because we're already navigating in darkness. Red-green reflections flame in the froth along our sides, and our only vision is the radar, whose empty green screen Matti watches over with watery eyes. My exhaustion is unrelenting. I shake myself. The direction of travel arrow on the compass is hard to see.

Keep Counting for a Minute More

Dark water roils beneath us. The boat gyrates wildly, jumping and thumping. Its motions are impossible to predict, like holding onto a furious wild horse. The compass reading swings from side to side. I try to discern a middle course by instinct alone. Matti is dictating directions in single degrees, and my compass only has a five-degree scale. I interpret, estimate, and try to hang on. A pernicious exhaustion has struck me, and I can't get rid of it.

"Now straight for minute . . . that's it . . . wait . . . left two degrees . . . straight, slow down, take it easy . . ."

His voice is deep and gravelly.

I am fighting at the wheel. I shake myself, forcing my eyes open. I shout, roaring myself awake, again and again. I nod and then tear the sleep away, just for moment . . . Drive, keep driving . . . to the harbor, two more hours . . . compass line . . . I can't see it . . . Pekka, don't fall asleep, you can't . . . to the harbor . . . keep going . . .

I'm driving in the dark, completely dependent on Matti's instructions. I know that we are surrounded by jagged, high crags, and I can feel the strength of the roiling current.

"Matti, give me an overview. I have to know!"

"To starboard is a wall one mile away, less space to port, and an opening ahead. That's where we're going . . ."

"OK. I'm beat, but I'm going to keep driving." I'm yelling now and then—I have to.

"I can see and hear . . . it's all the same out here . . . let's just keep going . . ."

". . . we . . . let's go . . . give it to me in fives, if you can . . ."

"Accelerate a little . . . good . . . hold here, three five seven."

"Three five seven."

The acceleration perks me up a bit and calms the compass down. The bottom bangs violently. I'm afraid for the propeller. I accelerate more, but I don't feel any shaking in the engine.

"What was it? Did we hit something?"

"I don't know. I don't feel anything in the prop at least . . . it was . . . yeah . . . The sea god Ahti whacked us with his hammer . . ."

"Yeah . . . that's what it was . . . Ahti having a go . . ."

The compass line has disappeared completely. Now I'm reading the middle course from the edge of the instrument dome. I know the shape of the gauge from years of use. I have to keep going somehow.

"Stop. Let's take our bearings."

I disengage the motor and feel how the current starts to move us and toss us again. The hull slams from side to side uncontrollably, the hits tearing at our innards and the rails on the bow dipping deep and then shooting up.

I reengage the engine, forcing the boat to stay in place, swinging the wheel back and forth.

"Good, you got it. Now hold there!"

I hold the shuddering boat in place, accelerating and pulling back according to the bucking of the water. My head falls, my eyes shut, and then I startle awake again. Still an hour. Stay awake, stay awake. Otherwise destruction!

Matti counts out loud. His speech is slowed, and it sounds like his voice is coming from a cave. I hear his struggle:

"There . . . about . . . should be . . . why can't I see . . . this isn't supposed to be . . . why are the radar and GPS giving different positions . . . hold on, don't let it, there, yeah . . . dammit no . . . how is it . . ."

I hear and understand, but I don't budge. I try to rest. My left hand turns the wheel and my right works the gas. My eye watches the edge of the compass dome, evaluating what it sees, and the bottom bounces beneath us. I hear that the locations given by the instruments are conflicting, and I understand the meaning of that, but I remain silent and rest. My questions wouldn't help Matti. They would only bother him. He is the navigator and he has a job. He'll work it out. I know how to interpret what he's seeing and then it will be my turn, my turn to drive this sheet-metal sled through the boiling stream. I just rest. I have to rest. Otherwise we will never make it to the harbor. Do your calculations, Matti, keep counting for a minute more. I trust you. I'll push the boat through. Even if you give me the course to hell I'll drive straight there. I trust you, and you trust me, but keep counting for a minute more. Let me rest, just half a minute . . .

The foggy harbor opens up ahead of us. A passenger car out on the point of the breakwater flashes its lights—for us. We reported in to the Shetland shore radio two hours ago; they knew we were coming. The *FinnFaster* pushes up alongside the high concrete pier at an idle. I back up a little and Matti throws the rope up. I turn off the engine. I hear an ear-splitting silence and sigh.

I climb the few meters up to the dock slowly on the narrow, rusted iron ladder. My brain pounding out commands: Slowly . . . carefully . . . stop in between . . . you won't fall . . . slowly . . . up . . .

I reach up onto the dock in my survival suit and for a second I'm on all fours on the ground, nearly trembling. My eyes rotate around slowly: Suddenly I'm far away in the English Middle Ages. A dense fog enshrouds low, gray, rough-hewn stone buildings. The wind is already howling loudly. The wires are chattering and a massive gull is screeching on the end of a pylon. My gaze comes to rest on the ancient bell tower at the edge of the square. Three bronze bongs ring out through the fog.

I'm Sorry

The strong wind brought with it a beating rain. Angry torrents shake the canopy, and the flag ropes howl and whine. I startle awake frequently, hearing the sounds of the traffic coming to life. I don't react in any way to the strange speech high above. The only important things are the proximity of a solid pier, the warmth of my sleeping bag, and the support of my reclined seat. The Frenchman's weather fax flits by at the edge of consciousness.

"I'm sorry."

The taxi driver spreads his hands—he doesn't know either. No one knows where we can get fuel. Everyone is in a hurry to get somewhere, and no one is interested in this little Finnish boat's need to refuel. They listen with polite aloofness, glance at their watches, and express their apologies courteously. We are confused. Is it impossible for them to understand that there is a small motorboat in the harbor that has come from far away and now needs six hundred liters of gasoline in order to continue on its way? We are civil and wish to pay with hard cash. I understand that there isn't a bunker station with gasoline at this dock—normally at sea you use diesel—but it has to be possible to fill up somehow. It has to be possible to arrange, and that starts with asking the locals for directions.

The office is closed for the moment. On the door is a prim sign that reads "Exxon Office." We leave a note on the door, including our business, our location, and how much we need. I'm relieved—this is their line of work. A glass of milk and a bread roll help me make it another few minutes.

A young man has opened the office. I go in feeling buoyant. My note has disappeared from the door. He looks at me distantly. I recount our problem again.

"I know, it said on the note," he answers.

"Isn't there some way . . . six hundred liters . . . transport somehow . . . I can pay with card or cash, however . . ."

"I'm sorry."

I can't believe my ears. How can this be so hard? Where am I going to ask if not at the oil company office? It's already late. The wind is dropping. The next stage of our journey is waiting.

"Listen, the boat is in the harbor. I can move it . . . wherever, just so long as we can fill up . . ."

"I'm sorry, but we can't help you."

"Well, can you even suggest something . . . ?"

A faint smile flashes across his lips. "You could carry it in cans."

I look at the young man for a long time. Is he having a joke at my expense? Can't he do anything? Why isn't he even trying? I grit my teeth and say, without blinking an eye, "It's always nice to travel to new places and find people so helpful. Thank you. When you come to Helsinki, get in touch. I'll do whatever I can to help." I offer him my card.

"I shall never come to Helsinki."

We're well into Friday afternoon. The drizzle is intensifying, making us wet on the outside as the sweat from tramping around is doing the same from the inside. My stomach is rumbling, but we have to make it in time. We have to solve this simplest problem in the world, soon. The weekend is coming, and that won't make anything easier. I don't know whether to cry or to laugh. I look at my reflection in a shop window, doubting myself. I see my wet cap, my wild beard, my wet windbreaker, and my wet, dingy jeans, as well as the dirty, squelching sneakers on my feet. Do I look untrustworthy? Is my language ability insufficient? Why is this problem impossible to solve? I don't believe that—there are other wet people here too, and my English is just fine.

I hail down a van, and an elderly man peers out.

"Would you drive to the harbor? I'll explain on the way, and I can pay."

I tell him about our problem, and he nods as he drives. I feel relief. We stop at the dock.

"Matti, we're in business. We can haul it with this van . . ."

"What van?"

I look back. The van is swinging onto the street.

The white plastered office building is close to the street, and there is a large storage yard in the back. A man is wiping down the headlights of a tanker truck. The yellow on green BP painted on the side feels safe and familiar, a relief, like the finish line for a marathoner.

"What's in your tank?"

My heart leaps. I knew it would pay to stick with it! A tanker truck with just what I need, gasoline, square meters of it, barrels of it.

I glance at my watch—less than half an hour to closing.

"Wait just a minute. Where's your boss, the highest person in charge here who can make decisions?"

I run up the narrow stairway to the top, take a deep breath, and step inside.

The boss is expressionless in his necktie. He doesn't want to have anything to do with it either. They aren't in the habit of driving down to the dock to refuel boats. It would be possible in practice, but it just isn't the way.

I am dumbfounded. What kind of a farce is this? I try one more time: ". . . your truck in the parking lot . . . six hundred liters . . . I'll pay right now . . . it's very important . . ."

He straightens his tie and glances at his watch. When he speaks, it is only with his mouth.

"We can't do anything. I'm sorry."

I explode: "Doesn't anyone do any thinking surrounded by all these goddamn traditions? Do you royalists really mean we have to run around your streets carrying gas cans for days on end? If we don't get any fuel we're going have to stay here and you'll regret it!"

"I'm sorry."

My outrage boils over. With all the venom I can muster, I recommend that Shetland never declare war on anyone even by accident. If this is how they get things done, things would go extremely badly for them. I leave, slamming the door for good measure.

I stand on the pavement, out of breath. I've been running the whole day and business hours on the island are over now. I am incredulous. What an unbelievable farce. Incomprehensible. A normal business transaction has first grown into a difficulty, then a problem, and finally a nightmare, but still people are moving about with their bags and briefcases and cars like in any other village. They always reply, but all they are is sorry.

I light a cigarette and think. We're going to fill our tanks and we're going to leave here, dammit! I'm done begging!

Here it comes! I jump out in front. The empty red truck stops, its brakes squealing.

I coax, pressure, and finally force the large vehicle to come along with me. The driver resists, glancing at his watch and muttering something. I slam the dashboard with my fists and motion at the street:

"That way, let's go, faster! This is urgent!"

He furrows his brow and glances at me out of the corner of his eye. He's flummoxed. I drum the dashboard with my fingers, my face serious.

"Let's move it. You're free once the boat fuel tanks are full. Then I pay. If you try to take off, I'll smash your face in."

In the darkening evening the boat finally gets her food. With the large truck we haul one of our deck barrels back and forth through the city between the service station and the harbor. We spend the rest of the evening filling the tanks with six hundred liters using two funnels and our own tube. At times we have to wait for future top British sailors to get their dinghies dragged to the dock. Their handling looks exasperatingly slack. It's getting close to nine o'clock.

A little fuel runs onto the side of the dock. A gentleman in a bowler hat points disapprovingly at the splotch of gas in the water with his cane. "I'm sorry," I answer slowly, feeling like tossing the old guy in the drink. The BP tanker truck is standing only a few hundred meters away as I pay almost one thousand marks to the truck driver for his services.

The Royal Yachting Association's small pub is almost full. We push our way in, our stomachs screaming and our mouths parched. The time is almost midnight, but our exhaustion is overcome by our knowledge of our full gas tanks and the moderate forecast for tomorrow.

The well-groomed men at the bar have on white shirts, bow ties, and plaid kilts. We greet them with clear voices and a bow. They do not reply or even acknowledge our presence. I am embarrassed for them. How can they be such boors?

I feel the tape recorder in my pocket . . . I get an idea . . . Could I do it? I'm curious . . . Let's see what happens. Cultural anthropology has always fascinated me . . . I'm going to do it!

"Excuse me, gentlemen. I am from Finland and here for the first time. I would like to do a program for the Finnish Broadcasting Company about charming Lerwick and especially about the cozy, sociable atmosphere of your pub . . ."

The hum of conversation dies; they look at my tape recorder. They start jostling each other. I hear whispers: ". . . it's going on the radio, a real radio program . . ."

A circle forms around me—everyone wants to talk. They clap me on the back and a row of full pints appears before me.

". . . this is a splendid place, very old and everyone is welcome . . ."

I hear their intoxication, especially with their own words. I sip through the battery of beers and let the recorder run on the table and them at the mouth. Eventually it all begins to disgust me, and I want to get away.

Our boat is waiting at the dock, and the bay is calm and foggy. The full load of fuel has raised the waterline a bit. I climb down the ladder, open the zipper, and crawl inside. A candle begins to warm the damp cockpit. I'm coming home from a foreign country where the steering wheels are on the left, no one understands Scandinavian, my mobile phone doesn't work, the money is difficult, and the people are sorry. But this is still their island, and we are only visiting. Before falling asleep I think about my call home. Our divorce papers haven't come yet. Elina and I were both silent for a long time. I am a strange wanderer far away and don't even know my marital status.

The Blue Hour

Our roiling wake draws a line away from Lerwick around midday. We laugh at the reserved Brits. They wouldn't come close. As the boat receded from the stone pier the group of people waving goodbye quickly grew. From a distance we see that there are rather a lot of them. Lerwick recedes, still a mystery.

We are rested and in motion. Matti looks up from his map.
"What did you pay for the shoes?"
"Just a couple of hundred marks. They felt sturdy. The girl was a real surprise—extremely helpful and not sorry even once."
"Aha, so she wasn't . . . that's strange . . . What are you reading?"
"Three-two-six."
"Bring it down a little, to three hundred and twenty."
"Three-two-zero."
The gasoline of the British Empire burns happily in our engine as we weave quickly through the island labyrinth of Shetland towards the open sea. Matti takes pleasure in his route choices—we are free. Not even the shipping lanes bind us. The high, grass-covered islands see the *FinnFaster* speeding towards the north-west, whence

we will move on another couple of hundred miles, to the Faeroe Islands.

An ancient lighthouse makes our imaginations soar. We make guesses about what the thoughts of the old lighthouse keeper would be like after seven, seventeen, or twenty-seven lonely years . . . We imagine that in the end he would begin to talk to his lighthouse, hating it and loving it, kicking it as he walked by, sulking at it for a week or two and then relenting once again to its warmth . . .

The open Atlantic offers trying, choppy seas. I'm forced to drop our speed to about ten knots and concentrate carefully for hours on end. A large creature dives alongside us, trying to hide. I don't have time to get a good look, but my retinas are left with an image of a walrus head. A greedy desire to make progress lashes me forward. I drive at the highest possible speed given the conditions. The bow railings curtsy in my field of vision, spray flies to the sides, the bottom thuds and thuds, and the motor bellows.

As we leave the shoreline behind, the swells grow, but calm at the same time. I'm able to increase our speed to fourteen knots, which lightens the mood. The thought of crawling two hundred miles across the ocean was not pleasant.

Work in the cockpit of a boat is based on routine. Your eye moves along its own, subconscious path: compass, log, RPM's, swell. The wheel searches for gentler slopes to climb, constantly adjusting course. The right hand sits on the gas lever, adding or pulling back as the situation demands. The thumb controls the position of the engine by adjusting the electronic trim. I use the gauges to search for the most efficient settings, the highest speed at the number of revolutions I've chosen, and I keep the boat on course using the compass and GPS. Matti tracks our course on his own screen and looks for any possible obstructions using the radar. During open legs he might even nap for a moment. Songs and stories fly as we try to push the inevitable exhaustion farther off into the future.

"It happened in Rochester," Matti begins. He goes on to tell of a mess-boy from rural Savo—Eastern Finland—who started feeling up a man in a kilt while drunk in a bar.

I almost can't stop laughing.

"What happened?"

"Well, the mess-boy flew out onto the street . . . and hard, goddammit! And, man, did that Scott yell! He was big . . . ugly as sin . . . a real bonehead . . .!"

The evening is clear, with only a few clouds, and dry. The sunset paints the horizon a bluish violet. The sea is calm, so I'm able to get up to twenty knots, and Matti is singing to himself. I sense my own excitement and the beauty of the open sea on the cusp of night. I rejoice in being so far away. I rejoice in how I feel right now, right here, deep in the bosom of the distant Atlantic. My thoughts fly home, to my sons, Elina, and Muru. I am at one with it all. This is good.

We stop to take our bearings, at the same time filling the motor's oil reservoir and chewing on some dry rations. We have to pay attention to our energy—the coolness of the air can eat away at it unnoticed, and hunger and low blood sugar are risk factors.

During the break I lower my backrest and relax with my eyes closed. We rock in the two-meter swell without talking, just resting. I crack my eyes and the blue-violet shimmering of the sun burns before me over the hazy silhouette of the Faeroe Islands. I don't feel any hurry to get anywhere. I am here, where it is good.

The knowledge that I will meet a new culture sometime in the early hours of the morning makes my heart race. I am curious to meet them, a few tens of thousands of people living in the northern heart of the Atlantic.

"It can't be true, goddammit, it just can't!" Matti is laughing, his face hairy, shaking with laughter in his survival suit, his eyes watering. He shakes his head, goes silent, and then explodes again. "It's true, every word."

I'm telling him about an incident in May in the yard outside my house. I was telling one of my neighbors about how I was casting off for Iceland in an open motorboat in a few weeks' time. He looked me in the eye for a long time and cocked his head silently. Then, obviously ill at ease, he drew in the dirt with the tip of his shoe and then observed, slowly, "So you won't be making it to the yard cleanup pitch-in day?"

The boat is gently loping along. I look at the bow railings. They have become so familiar, bowing and rising before us for days on end. I wanted railings shaped a certain way, the kind I've always used. The railings on the *FinnFaster* are very high, so that tall men like us can get good support from them in high tidal docks when we're reaching up. Juha understood the idea as well and built them accordingly.

Another three hours have passed, and we are taking another break. It feels magnificent—I can move around freely. I've turned off the engine, and the only sounds are the splashing of the waves and our infrequent speech. Matti is calculating our position—his chatter with himself feels somehow homey. I feel good as I listen to his one-sided conversation. For a long time now our bond has not only been the bond we feel as a team, the relationship of navigator and driver, but much more. We are alone and far from home, and we are headed even farther away. Our lives, our values, our troubles, and our joys have been laid bare to each other. We value each other's contributions and know that alone neither of us would survive. In order for us both to make it through we also support each other emotionally, unnoticed, without paying it any heed, truly.

We let the break stretch out, talking infrequently, rocking and resting. We can see the Faeroes, and the weather will surely hold, so we enjoy this time. This is the blue hour, with everything in balance. We are far away, but we are ourselves. We both feel good. Everything is in order externally and internally. We have been driving for a long time, more than around the clock, and the hours on our motor are already approaching one hundred and fifty since we left Helsinki Market Square. Soon we start back up and pull the rest of the way in. An hour before reaching the harbor I see the first whale of my life.

Storm Island Maija

In the middle of the Atlantic, the beautiful Faeroe Island morning augurs the imminent rising of the sun. We drive straight into the colorful heart of Tórshavn. Again we are reaching a harbor at night, again a new dock and stretching of our limbs. Again we breathe in the new environment while people sleep in their homes.

Four young people wandering in the early hours rush to our boat. They have with them a bottle of booze, a guitar, and friendly, bright eyes. We shake hands, and I feel moved by their openness and how full of life they are. They are overjoyed at our arrival and wish to help however they can.

Matti borrows the guitar. He sits down on one of the pier posts and checks the tuning. We are tired, but we feel good. "Storm Island Maija" begins ringing out on that distant island dock. I watch my friend—he closes his eyes and relaxes as the guitar reverberates and his fingers loosen up. He strokes and lashes the instrument in turns, filling the Tórshavn dock with the tender melody. The last time he played was at home in Vihti. The youths listen in silence. A tired Finnish traveler in an ocean survival suit on his way from somewhere to somewhere is playing right now for the whole world. I listen and watch in silence. I feel his music and want to share the sensations he is experiencing. We are friends and we trust each other. Right now this guitar is important.

Matti unzips his survival suit and continues playing. Right now this is what he wants to do. We aren't in a hurry to be anywhere. We just reached harbor again, the motor is cooling silently, and the flags of the *FinnFaster* are flapping in the rising morning wind. The weather is cool and the orange of the sun is beginning to color the wall of the seamen's home.

The Column

Being in the Faeroes is nice because of the people. This miniature society only about the size of Hämeenlinna, some 50,000 souls nestled in the bosom of the vast Atlantic, is friendly. I can feel it in their eyes, their handshakes, and their smiles. Their life is not idyllic—the ocean around them makes sure of that. The directness and openness is the same I have met before in the middle of the sea, on Magerøya off Nordkapp, on Kökar, and now here. Their spiritual connection to Scandinavia is strong; we are friends and our Norden baton is an important symbol. As people who know the sea, they value travelers from afar.

A car pulls onto the dock and a friendly, dark-haired man comes up to the boat with a notebook.

"How much do you need, what kind, and when can we bring it? Just go ahead and tell me. I can come any time that works for you. You need rest. You must be tired, just tell me . . . All the way from Finland!

We arrange a refueling time that works for everyone.

"Nine hundred liters. I'll be here then . . . That's a lot . . . 1,200 all together . . . What a trip . . . I've been to Iceland once . . . All the way from Finland . . . you are friends . . ."

These eyes and hands, their genuineness and directness, are touching and exhilarating. I often think that I could live here, easily.

The harbor is sleeping; Tórshavn is resting. The summer night's sun wanders for a few hours below the horizon of the Atlantic. The light is a bright brilliance—it isn't coming from anywhere—an all-illuminating whiteness.

The high-bowed, colorful Faroese boats nod drowsily alongside us in their moorings. I look all around at the bow shapes of the boats and see in them the sea. Authentic, traditional bows follow the forms of the sea. They are made for their element—their purpose is to meet the waves. Here the waves are high and steep and the bows are like on Viking boats.

The headfast on the *Ritan* next to us creaks. The five by ten meter, black and white ferry boat also has a high bow; it needs it throughout the year in the roiling breakers along the shore. The black-haired skipper of the *Ritan* does not talk much, but comes over to our boat often. He is helping too with his silent presence.

I am inspecting my notes. Matti has fallen asleep. A light snoring fills the cockpit; he tries to roll over, but the seat is not a bed. In a few weeks we have learned to sleep on the seats just with the backrests laid down.

A telephone conversation floods into my mind. My eyes come to rest on a tern sitting on a pylon, my friend.

"You don't have to say anything more, Pekka, I know . . . I just do . . ."

That's what she said after we had talked quietly for a long time. I could feel her closeness, her breath, her caress . . . Muru shares my sensation. She feels the same, she says. She understands. We don't need many words. I felt a strong arousal, which began from a feeling of joy. I told her without shame, quietly, and she forbade me from continuing. She knew; she felt the same.

The spirit stove had gone out long ago. I feel the bottle. It is cold. I could fill it a little. It would dry the moisture for a while. I open the top and pour some liquid in. I dig out the matches, but I don't have time to light one.

WHOOOSH!!! A violent flame makes a direct hit on my face.

I dodge and bend over, smelling the stench of burning hair and beard. I open my eyes.

This can't be true, it can't . . . Dear God. A fire. Everything will burn. . . .

The canopy over the console has caught fire. The liquid that splashed from the bottle when I jumped away from the flames has ignited. The fire spreads quickly. Half-meter flames are blazing on the floor of the cockpit in a second.

"Matti, Matti, wake up. We're on fire . . . !"

There was a splash on the footbox of Matti's sleeping bag, and it also bursts into flames. Matti doesn't wake up. I slap his cheek and try to stomp out the floor.

"WAKE UP, MATTI . . . FIRE!"

"WHATWHATWHAT, FIRE, WHERE? WATER, QUICK, WATER!"

"NO WATER NOW, RUN, MATTI! RUN AWAY!"

I kick Matti's sleeping bag out and snatch the knife from my station. With a few strokes I rip the canopy free and fling the burning pieces into the sea. I rush to the fire extinguisher . . . quick, quick, the fire is spreading . . . There are gas tanks below. We have a lot of gas . . . There is a plastic canister on the bow—the fire is close . . .

. . . Goddamn fire extinguisher, won't come loose. It was a mistake to fasten it with a strap. The fastener is rusted shut . . . won't open . . . the fire is blazing . . .

I jump into the middle of it, not thinking. I bat at it with bare hands, stamping it, kicking it.

The sheet metal opens my palms, blood gushing with every stroke. It burns. I hit, I pound, I drum, I jump, I kick all around, ahead, to the side, behind, whirling around . . . quick, quick . . . not much time . . .

I ask myself how much time I have left to live. One second or five? I don't know. I'm going to try to beat it. Below there are 150 liters of gasoline and 650 liters of explosive gas. Just one spark, just the smallest spark would be enough to explode the entire ship, everything, the whole dock, into smithereens.

The service hatch we cut on Gothenburg. It has tar seams, and beneath it are the tanks. I hit the burning tar. No burning drips can fall below!

What to do? Run or hit? By running I might still have time to get out of the middle of the terrible explosion. Do I have time to beat it out? How much time do I have?

My hands pound away, bloody. The bottoms have burned off my wool socks, and the soles of my feet are stinging, but I'm not thinking about anything but the impending explosion. I have to stop it!

My beating binds me in place . . . this flame, that flame . . . suddenly over there . . . that one is licking the seat . . . quick, over there . . .

The flames begin to subside. In the end I can't see any more. I stand there panting and look at the extinguished deck carpet. My nightmare, THE BIG EXPLOSION, slides past my eyes slowly and a paralyzing wave of relief rolls over me. We are alive and we still have the *Finn-Faster*.

We talk quietly, already well into the small hours. We have been given a gift—our trip has not been cut off. I see the column, the enormous column of flame and smoke where our journey would have ended. The column would have destroyed the pier, the *Ritan* and dozens of other boats in the harbor, and my life would have ended. The column did not come, though. It was not its time. I inspect the seam of the service hatch.

"Matti, some pieces of this tar seam have burned off completely. There were drips."

"I was looking at the same thing. The tanks are sealed tight." We glance at each other. No questions, no answers, but we are thinking the same thing.

Drips of burning tar made it onto the tanks, but there wasn't any gas because Timo Rokka's welds didn't leak. It would have taken only one gas molecule, but there weren't any floating down there and we will never know how close we came.

"Your hands look pretty bad. Can you still drive?"

"Yeah, sure. Of course. I'll probably have to go to the barber though," I answer, seeing in my mind the column of flame that never was.

The Fog is my Friend

A slender, blond woman pours our coffee. We have been talking for a long time now, Finnish and Scandinavian all mixed together. We suddenly have a lot to say and listen to each other carefully. She looks me in the eyes and says calmly, "I go into the fog when I want to be alone to collect myself. If I can't see anyone else then no one can see me . . . it gives me strength, and I know who I am again. The fog is my friend."

I look at Tita. A little girl, sent to Denmark during the war, lost all those years ago. She doesn't know who she is or where she is from. Her mother—somewhere far away in Finland. What does that mean?

"I weave fog. It has so many forms—look at this violet. Look at the long, fluffy strands . . . look from the side. Do you like it?"

The little girl is sent to school in Finland. It is hard for her to speak; when she's rushed Danish words come out and her schoolmates laugh. The little girl with the braids cries and wants to go back to Denmark.

"Wools are so different. I love them. Now I'm using red because I want to encourage people."

The little girl grows into a teenager and then a woman. She finds herself and her life here on this distant island in the Atlantic. Here she creates her textile fog and is no longer lost.

I feel the warmth of her home, and it moves me. I had this too . . . long ago . . . was it . . . where is it . . . ?

Tita listens quietly. I dare to speak because she listens and wants to understand. The importance of fog to Tita shakes me, making me open up.

". . . the open sea and the fog, complete emptiness, nothing unnecessary . . . it's so solemn and immense . . . the grayness and a lone boat, nothing extra . . . my mass, my sacrament . . . The beginning of the Bible, in the beginning only the sky and the sea . . . I am safe there . . ."

"The fog is your friend too, Pekka. I can hear that."

Late that night I walk along the pier, deep in thought. Once again a tiny meeting had happened in the cosmos, in the universe, the microscopic paths of two people finding each other for a moment. They had important things to say and wanted to listen to each other.

A thick fog has risen over the harbor bay of Tórshavn.

He Rules the Air and Elements

A narrow rock stairway leads up to the second story. The walls smell of the dampness of a crypt and the steps bear the impressions of many wanderers. The paint disappeared from the iron railing ages ago.

The seamen's home is peaceful and simple. The tablecloths in the small dining room are clean; the lamb stew is fragrant, the cutlery clinks, and the diners speak in hushed tones. The boys in the corner office are helpful, changing money, selling postage stamps, forwarding letters, payments, and messages from the four corners of the world to the four corners of the world.

In the reading corner sits a group—we nod at each other, already familiar just by sight. We are friends in every sense of the word, they from somewhere and we from Finland. Our sweaters announce it. Here the peace of friendship, rest and assistance prevails. I sense a mysterious, mutual agreement in the air.

"Look, Matti. There's a harmonium in the corner. We used to have one at home a long time ago . . . I'd like to . . ."

I open the cover and finger the keys. I carefully play the melody of "Besame mucho." An old sailor in a gray sweater smiles and nods. I start humming, and Matti joins in. We start singing, gaining courage and continuing, switching to "Summertime," trying to find our harmony.

"Let's do something serious too, Matti."

"OK. How about 'The Spirit of Truth?' I like it; it reaches for something . . . incorruptible . . ."

The rock-walled room echoes as we sing loudly. Matti's baritone and my bass ring out together. The harmonium is in tune, and our souls open up. Small groups come in to listen. The young man from the office comes and stands by us, smiling.

"Keep going, keep going. It sounds great. It isn't often . . ."

We loosen up even more, seeing that we are singing for so many others as well and we've already been practicing together for nearly two thousand miles.

"Matti, it's your turn to suggest something."

"'We Praise Thee, Our Creator.'"

The room reverberates, adding to the harmony. Our singing becomes a mystical release, giving us strength. We glance at each other and continue. I bring the harmonium up to forte. The moment becomes real and powerful. The man in the gray sweater wipes the corner of his eye, and everyone listens quietly.

". . . He rules the air and elements . . ."

We are so far away. That night we start on our final leg, heading off for what I have been planning for years. Nothing is certain, and we will be alone out there. Our voices ring out—we have a majestic strength in our singing as we let everything roll out, unfettered . . .

". . . and calms the raging sea . . ."

I think of the words, and as I look at the devotional picture on the wall, violent chills run down my back. Matti singes next to me in his faded sweater. I think of the waves, the huge, rolling, thundering waves of the ocean . . . I see us from a distance: Two wanderers in the middle of the ocean happen upon a harmonium in a modest seamen's home . . . we are leaving the next night . . . what after that? A huge leg to cross . . . this strange journey, where everything is on our shoulders . . . might we be singing our own funeral dirge? No one knows . . .

We sing all of the verses peacefully, but seriously, with all of the masculine strength and will that has thrown us to this far-away room. We end, serious, a little confused. Someone nods, another thanks us quietly, and a few squeeze our hands. This is a surprising moment of unity—we are all far from somewhere.

We walk to the boat in silence. I think of the chills and the words of the hymn, the whole situation. It's just the two of us walking here, Matti and Pekka, strong and imperfect. We sang on instead to the end, every verse, and then we were ready for our departure. Was it our funeral dirge? Or simply a silent prayer? Where did it come from?

Haven't We Already Said Everything?

The evening sky over Tórshavn is reddening and the clouds are forming regular patterns as the colorful walls of the small homes are illuminated by the peaceful sunset. We are preparing to leave before midnight. A few more specific weather maps just to be sure, and then we'll be ready.

The hotel runner brought us an envelope. It contained an important fax from Finland. My friend Kjell Holm had been in contact with his acquaintance at a weather station in Ireland and sent us important information. There were moderate winds awaiting us and at least one low pressure front we would have to go through on this last and longest leg of our journey.

We have prepared the boat for departure, putting everything in its place, lashing things down and protecting what needs protecting. Matti is in town replenishing our food reserves. Crackers, chocolate, raisins, nuts and soft drinks.

I look at the sky. I am here on the final stretch of my dream, this journey I've been trying to foresee in my mind for years now. Despite the intensity of our work, I feel earnest, solemn. This is true, still. We really have made it this far from home in this small boat. Soon the longest boat trip of my life will begin. The unknown before me condenses in my mind as a seriousness, especially since this section is the longest and the last.

Even though we are so far away, we still have not made our journey. It would be dangerous to be lulled into an attitude that "we've got this" in the middle of our journey, prematurely. Every leg has to be approached with humility, with concentration and complete preparation for severe difficulties.

The seamen's home comes to mind again. Had my chills been some sort of premonition, or was I just feeling the extreme seriousness of the situation. "He rules the air and elements and clams the raging sea . . ." It is precisely these North Atlantic waves that I have been thinking about, researching, and doing calculations on for the past years. I have sat in front of weather maps for hours on end, calculating fuel efficiencies, drive times, propeller options—in a word, everything. The dozens of hours before me will be nature's test of all my preparations. The sea is incorruptible.

After a few days we will have the real answer. And what if, when that day comes, despite it all, we have been washed into the Atlantic, and what if no one has heard our distress signal or they aren't in time to find us? What if it was our funeral dirge . . . what then?

I quickly think of Elina. Would we be ready for it? I grab the telephone. The connection is fair.

"Hi, it's Pekka. We're starting the long leg soon. I wanted to call one last time. Just to be sure."

"Hopefully everything will go well."

"Don't you have anything else to say?"

"What on earth do you want? Haven't we already said everything?"

"What? Yes. Of course, we've said everything, every last damn thing! Yeah, we've goddamn said it all. Fine. Everything!"

I slam down the phone in rage. Oh hell! I try to call. I try to make it right, to ask forgiveness, to be forgiven, to give forgiveness, to understand, to be understood—anything before the most extreme trial of my journey. And the answer I get is "haven't we already said everything?" I curse and rage. I light a cigarette, then another, then a third. Doesn't this start mean anything more to her than that? Doesn't thirty years together have more weight than that, even now?

I sit in my driving seat, thinking. If it was a misunderstanding, what demon is it that always heaps them on us? I'm suddenly incredibly tired. There isn't anything new here, even though this conversation did involve more distance separating us and a more extreme backdrop than before. "Hopefully everything will go well"—hell, was there any more formal way to say that?

The boat is anchored firmly to the side of the dock. The compass needle is resting steady at 273 degrees, the echo is showing five meters of water, and the log is at zero knots. We are still a part of the working world, just like any boat in any dock. Tonight things are going to start happening! The compass will hover around 300 degrees for 20-40 hours, with the engine howling for just as long. The Atlantic will surge beneath us at nearly one kilometer of depth. Our eyes will be bloodshot and our limbs will go dumb. We have to be able to drive to our destination or make it back here. There will be no way to interrupt the journey—we have to get out of the emptiness to somewhere. The sea does not listen to explanations.

I'm still thinking about the phone conversation. What would I have said? I would have reminisced about the joys of our sons' births, a summer night on Kökar with our little Vikings already sleeping in the tent after a day of treasure hunting, our struggles, our homes, taking big risks, pressures, joys—our whole life. That is what I would talk about right now. Even though as the skipper of the *FinnFaster* I'm ready to turn the starter key right now, as a person I feel a deep powerlessness. How have we already said everything? Is that possible in real life? So am I supposed to forget or silence myself? And what if I can't do that? I can't talk about these things with anyone else.

Matti arrives. The result of our quick meeting is clear: We are leaving forthwith. We begin dressing in our driving clothes one more time. First a warm layer. Jeans and a wool sweater, in the middle the "bear suit", the survival suit's warm, stitched coveralls, a wind breaker, and finally on top the survival suit itself with its watertight cuffs, neck opening, and zipper. We are in our zone again.

We go through the checklist: latches, lashings, ropes, motor oil, drinking water, food, life raft fasteners, emergency beacon . . . everything we think could be important in different situations. This is exactly why there can be nothing extra in a small boat—everything has to be in order and under control. Cramped chaos is impossible to manage, and even less so on the Atlantic.

The final well-wishers come to the dock to send us off. A man in a clean-cut suit and dark skipper's hat comes over to talk to me. He is extremely purposeful and competent. He does not ask about our trip or our equipment, but rather about our route out to the open sea—he wants to give us advice.

There is a hint of tension about him.

"Matti is in the boat. He's the navigator. Tell him directly."

I feel a sense of satisfaction—advice based on local knowledge is always important, and we are in difficult waters now.

I start the engine, and we slowly draw away from the dock. There Matti is again at his familiar workstation in front of the radar, GPS, and maps. The clock says 2100. The final leg of our journey has begun. I increase the RPMs little by little, trimming the position of the engine and starting once again to settle into another long stretch of driving the *FinnFaster*. Our world once again shrinks to seven by two and a half meters, but our spiritual world stretches to infinity. We are free, which is good for the seafarer.

We begin to settle into our routine, even though the pause of a few days did feel like an interruption. This time leaving has an entirely different flavor, something special. In the cockpit we know without saying it that this leg is the last, the longest, and the most demanding.

The Atlantic night is fading to a dark twilight. A flock of terns wheels over us for a moment and then the harbor is already far behind. We are alone again in our familiar miniature world. The engine growls in the back in its familiar way, the waves are outlined as rhythmic lines of foam and spray far to either side, the bow dips, and the hull bangs below us. Rotations, more rotations . . . The banging of the hull quickens to a steady thumping as I lift the boat into a powerful hydroplane. The familiar bow railings curtsy in front of us. I trim them into place, and we're off. Matti reads off our route out to the northwestern border of the Faroe Islands, out towards Mykines Island. The grand finale has begun.

Mykines

"Matti, look out—LOOK OUT!"

I raise my arm to my eyes instinctively, to protect them from shards of windshield hitting my face, from the compasses and all of the equipment on the console, from the whole explosion. An express train washes over us, meters of foaming Atlantic exploding over us. I crouch down. The *FinnFaster* shakes violently.

I can't believe my eyes—the windshields are intact; everything is in its place. The frothing water tugs at our shins. We're swimming deep, but the engine is running.

"Hold on, turning!"

Have to empty the boat, fast! I accelerate into the next oncoming slope at full throttle and a moment before the breaking crest I pull the gas back, turn the wheel full left, and we're going with the wave. The extra-large drain ports in the rear bolt down the mass of water effectively. We're going back where we came from.

What is happening here? Before us, on the route we came a couple of hours ago out towards Mykines, the sea is boiling. It is impossible to follow the waves with your eye—there aren't waves, only a churning, ghostly hell come out of nowhere in an instant in the darkness of night.

. . . This can't be true . . . that's going to fall on us . . . I dodge, getting us on the crest . . . are we going to fall? I balance the engine, reversing for a moment . . . carefully . . . we continue . . . two knots on the log . . . don't think about it. . . just hold on . . . how far to that bad strait, Matti? 8 miles . . . that's a lot . . . four more hours of this . . . then much worse . . . Muru, where are you . . . your hand . . . shall we dance . . . Elina, the boys . . . I'm here . . . hard . . . damn . . . I can do it . . . I have to . . . we're going on . . . we're crawling along, in secret, we haven't flown yet . . . no mistakes, not one . . . just drive . . . now . . . is there a God . . . my mouth is dry . . . Matti . . . water . . . straight in my mouth . . . I can't let go . . . a green gorge . . . foam above . . . churning . . . driving a boat . . . we sneak out . . . I won't give up . . . titles don't help, only hands . . . the bow railings . . . up and down . . . we're falling . . . no . . . I brake, reverse . . . now fast forward for a second . . . Eetu, you can do it . . . don't give up . . . keep pushing back there . . . I'll guide you . . . I'll show you where to go . . . don't stop . . . goddamn, I'm watching you waves . . . you won't get us . . . I'm faster . . . still . . . can I do it . . . don't ask . . . I know you—you can't surprise me . . . hahaha . . . you missed . . . still hours to the strait . . . I am a driving machine, a driving machine, and driving machine . . . keep going, keep going . . . there isn't anything else . . . hours, hours . . . are you afraid, Matti? . . . I glance . . . no fear . . . you can't have fear here or you disappear . . . Matti is biting his moustache . . . good . . . light me a cigarette . . . straight into my mouth . . . I can't let go . . . I smoke, as if . . . the skipper is bluffing . . . no problem . . . go ahead and suck your mustache . . . we're just driving here . . . Matti is as cool as a cucumber . . . good . . . describes the strait to me from the radar . . . giving me headings . . . the navigator's work . . . I can drive . . . in peace . . . concentrate . . . Martti . . . my father . . . in the war . . . anti-aircraft firing . . . flies a loop, two, three, under fire . . . they laugh . . . that is difficult . . . accelerate and reverse, then turn that way . . . hahaha . . . we're going on . . . hell . . . death . . . pulling at our doorknob . . . you won't come here . . . you can't get in . . . I'm pulling against you . . . with my hands . . . myself . . . we won't leave, goddammit . . . not without a fight . . . we won't leave . . . drive . . . there, high . . . no, there . . . all high . . . the strait is getting closer . . . don't let it fall . . . carefully . . . Matti picks his nose . . . good, nerves holding . . . picking his nose . . . me with a cig smoldering . . . Lord, let us . . . waves . . . a truly rough sea, waves the size of mountains, real . . . breaking crests, take this away . . . keep your nerve, Matti . . . we can take it together . . . if you scream, I'll knock you senseless . . . I won't leave you, I'll just hit you hard . . . Pirkko, the children . . . I can't hit him, I can't take my hands off the wheel . . . Matti looks, picks his nose . . . good . . . the gap is coming up, it moves . . . look carefully . . . I'm not coming there . . . don't think, don't wait . . . you won't get us . . . reverse . . . hanging on the face . . . we're upright . . .

I'm not coming there, I'm waiting here . . . I'm more clever . . . I know . . . wave, gap . . . you have to move . . . soon you will move, and then I will come . . . accelerate . . . haha, you didn't get me . . . on we go . . . Juha and I agreed . . . hydraulic steering . . . it will hold . . . good . . . Matti . . . we . . . bad strait . . . soon behind us . . .

A light wave slaps the stern, and the world howls with silence. I have turned off the engine for a moment. The terns are screeching somewhere. A large truck fires up somewhere. The familiar harbor of Tórshavn lies a mile ahead of us as the morning fog dissipates. Six o'clock. Some early-riser parks his car on the dock.

"This is our decision given the situation. Now we have time to think." I put the NMT receiver back in its cradle—Finnish Broadcasting received their report, a matter-of-fact announcement that we were returning to our port of departure and not much else. All of my movements are slowed-down; time has stopped. The words I dictated in my report were logical, every period, comma, and pause carefully thought out, despite the fact that my soul was still in hell, a driving machine with teeth gritted and muscles tensed.

The waves lap against the aluminum hull. We say nothing. I light a cigarette, and Matti folds up the navigation charts and stows them in the console.

"That was a tight spot, by Beelzebub! Pekka, you suckered them every time. I saw it myself."

"What do you mean?"

"The peaks and troughs . . . you dodged them, every one. I saw it all."

"Matti . . . I'm wiped out . . . That took nine hours."

"Let's hit the sack."

We tied up at the same mooring we had left from in the evening. The skipper of the *Ritan* wondered why we were back.

"We couldn't get through."

"Which way did you go?"

"Mykines."

"Not possible!"

The bridge of the *Ritan* is clean and orderly. Our chart is spread out on the table, and the first mate hands us hot cups of coffee. We inspect our route of the previous night. The captain and first mate look at each other.

"Can it really be true that they made it back from there!"

We have just heard that the Mykines route is impossible: You can't go there. No one goes there. No boat has ever come back from there. You can't go there because it's hell during low tide with the northwesters.

I sit behind the wheel of our boat and look at the compass, lost in thought. I can't believe that our windshield is still in one piece and everything is in its place. Was that night really real? Exhaustion trembles in my muscles.

Matti is walking slowly around the dock with the zipper of his survival suit open. His brow is deeply furrowed, and his hands are behind his back. Now and then he turns, changing direction and continuing his pacing.

I know what he's thinking about.

I'm thinking about it too. We don't know the answer, but here we are at the dock.

My body, my eyes, my whole soul are screaming for sleep, but I can't rest. I was lashed to this spot driving for nearly half a day with my concentration stretched to the limit. The wheel spun without rest, the throttle flying forward and back, my thumb adjusting the electric trim at split second intervals. My eyes scanned the surging, green jaws of the ghostly mass for a path back to this dock.

A muscle cramp gives way to trembling. The scale of life has changed. I don't think about going on—or anything else. I am still locked in that same concentration. I continue the fight for a long time, even though the dock is cement and the knots are holding.

At seven in the morning I call Muru.

Her sweet voice rings familiarly in my ears, and I can see her smiling in her dress. I open the stout zipper of my survival suit and close my eyes. I am sitting in the driver's seat of the *FinnFaster*, standing at a dark gate and enveloped in Muru's arms all at the same time.

"I'm still alive," I get out, my mouth dry.

Matti is already snoring. I drink two glasses of vodka, lower my seat, and fall unconscious in a second.

The Interrogation

I don't snap to until I'm on the dock. I'm in a stupor and can't understand anything. Two plain-clothes officers display their badges and ask my name.

I didn't sleep, I just drowsed in a sort of haze. Waves washed over me; there was only a roiling darkness. My shoulders are on fire, and my arms ache. I heard the knocking for a long time, and finally they banged on the side of the boat while someone yelled, asking for the skipper of the *FinnFaster*.

"Your route, your radio traffic last night? What time?"

One of the men writes down my answers, and I begin to become distressed. It's like they're interrogating me. They're very formal.

"Your precise route and VHF radio messages?"

What's going on? What have we done? This is an interrogation. What have I done? Fought the fight of my life, driven the *FinnFaster* back to the harbor, struggling and pulling, using all my tricks and inventing a whole pile of new ones. Why am I being interrogated? What has Matti done? Been at my side the whole time, encouraged me, forced to sit calmly in the middle of a boiling hell. Our few conversations were calm, and his placidity helped me. Anyone else would have screamed, and I wouldn't have been able to concentrate. We needed all we got. I'm tired. We aren't criminals. Our radio traffic? Our radio was cold the whole time—I know that—but I'm not going to tell you, because I'm afraid of trouble. Yes, we have an emergency radio. We didn't use it . . .

"We didn't even touch the radio! We just got out of there!"

A moped accelerates up the street. A fisherman looks at us for a long time as he passes by. I am like an aggressive lion, ready to tear anyone to pieces who threatens us. Matti has to get to sleep, as I do as well. We are alive after all, and I don't know what you're after!

"I demand to know what this is all about. Why are you interrogating me and making notes?"

The men look at each other. We start conversing more normally. They are from the Faroe Island radio station, the chief and an assistant. Sometime past midnight the station received a vague report that a small Finnish boat had issued a distress signal around Mykines and that the connection had then broken.

I am dumbfounded. "Distress signal, small Finnish boat, Mykines, connection broken . . . no—what are you talking about? We didn't make any radio connections at all . . . I don't understand . . ."

We don't have anything to do with it. I'm too tired to even think about the whole mystery. Our meeting ends, and they shake my hand warmly.

"We're sorry for the disturbance. Try to get back to sleep. You're obviously tired. You must be."

I crawl back into my sleeping back, confused but unable to think about anything. Matti has woken up.

"This is really strange, Matti."

"I heard. I woke up when they started banging and heard the whole thing."

We sleep four hours and then eat. Rain drizzles down. I'm still tired. We go to a cafe; the black coffee is an elixir.

"Pekka, listen. That guy."

"Who?"

"The one in the gray suit with the skipper's cap. From the other night."

"Matti . . . goddamn it!"

"He did this. He sent us on that death route and then sent the message just to be sure! He must have known we wouldn't come out of there—even though we did. There isn't any other possibility. You heard what the *Ritan's* captain said, and he has to know what he's talking about."

"Well, of course. Distress signals don't send themselves, and in a place as small as this there can't be any uncertainty about which spots are passable.

And that guy didn't have any other business but our route when he came alongside. That was all he was interested in."

"So much for that. I am going to call Suni, Tita's husband, at some point though. The police need to know about this."

As shocking as it is, the knowledge of this attempt at sabotage calms me—I can understand the nightmare now. We were in the worst possible spot but we made it through. We begin assembling the latest forecasts, fill the tanks again, and move to the edge of the Vestmanna Fjord to wait for night, according to Matti's original plan. We are ready for a new grand finale. We point our bow towards Höfn, and continue on our way.

Location Infinity

Something beeps far away. I pull my sleeping bag up over my ears and continue sleeping. The beeping grows louder, more insistent, coming closer and disturbing my rest. I regain consciousness slowly, not knowing where I am.

. . . the boat, my driver's seat . . . the dock . . . darkness . . . Matti rolls over . . . a white van above . . . a light and . . . the clock beeping loudly on the console . . .

"Wake up, Matti. It's two o'clock—time to cast off." The Vestmanna fishing harbor is asleep around us.

We drove into the cove of the fjord in the evening, up next to the fish factory, to sleep and wait for low tide, to use it as we head out into the Atlantic. The route is the one we originally chose. The previous night's odyssey disrupted our plans.

A smiling man gets out of the van. He has two beautiful salmon in a box.

"As a gift to you, to take with you."

The darkness and the sleeping village make us lower our voices. We talk quietly for a long time under the lamp. I realize that he has been waiting next to us for four hours with his gift; he didn't want to wake us, because he knew how much we have ahead of us. The surprise friendship of this complete stranger gladdens my heart.

"How long will it take you to get there?"

"It's hard to say—maybe a few dozen hours. It depends on so many things . . . the swell . . . our energy . . . there are low pressure fronts to go through . . . how they hits us . . . everything is in motion . . ."

"It's a lot to deal with. And you'll get tired. No radio contact either. You can't get lost; navigation is important . . ."

"No, we can't get lost. And we never have once. My pal's been shooting bull's-eyes the whole trip, night and day in fog as thick as oatmeal, no matter how tired he is. He's a sea captain . . . good . . . he always counts out loud . . . yes, we get along . . ."

We start freshening up, doing calisthenics, walking a little, and organizing our small supplies. We eat a light meal without talking much at all. We slowly charge our batteries, side by side. The gloom has lightened a bit, and the dark bay surrounded by green mountains is calm. The chain of white houses that winds along the slopes is dark.

We check the deck barrels. We have made them fast with strong cargo straps, several times over. They could come loose when we are thrown up and then come down like a bomb on our heads. Everything is in order.

Matti marks our departure in the log book, we nod to each other, and I turn the key. The engine grunts rapidly and belches a small blue cloud. We slowly push back from the dock and wave to our friend. We idle past a large trawler and then slowly leave sleeping Vestmanna behind.

"She's really blowing. Good God!"

We are already far off near the mouth of the fjord, and the strong wind familiar from Mykines is whipping around us. The confused water surface is what causes it. Again the current and the swell collide, the ocean struggling with itself and the surface of the water being rent into vortices or angry, high-lashing tongues of water. I am concentrating, fighting my stress. A repeat of last night?

The churning grows, so I have to lower our speed, but I feel that I am in charge. The boat is in hand and we are progressing slowly in our strange dance towards the mouth of the fjord.

As the cloudy morning dawns, we push out into the foggy open sea. A memory of an early morning departure from Porkkala in Finland on a salmon run flashes through my mind. The waters of the fjord are still moving beneath us, but less so, and I increase speed. The banging on the hull increases, and the bow strikes the peaks rhythmically. We begin to be tossed by the growing waves as the drizzling fog thickens and the surging grayness deepens. Höfn is directly before us, but three hundred miles away, with the North Atlantic between us.

Watching my driving irritates me. I'm taking the waves awkwardly, mechanically. I'm not completely here; something is missing. I consider whether I'm nervous, and I don't know. Is the thought of the length of this leg making me tired? We've estimated the duration at something between twenty and thirty hours of straight driving. I can't figure myself out. I just keep driving, hour after hour, but I've lost my touch. It bothers me. We don't talk. We just move forward, not seeing anyone.

"Take it a little to port. Keep her at three hundred."

"Yeah . . . got a little past . . . three hundred."

Holding on course becomes more difficult. I try to pull it together, but the compass reading wanders constantly; my eyelids are sagging, and the magnitude of the job ahead worries me.

"Matti, let's take a break! I'm really tired, and it's hard to keep the heading."

"I noticed . . . let's go one more hour. So many more miles . . ."

"OK, I'll try, but I don't know what's going on. It just isn't working . . ."

We continue on, but driving becomes impossible. Time shortens to agonizing minutes of battle, constant pitching and corrections. I can't do it.

"I'm stopping now. No choice!"

I shut off the engine and we rock in the midst of the gray fog. I lean back in my seat, feeling the oblivion of rest. Waves squelch from the stern into my consciousness, and I listen at ease to the story they tell.

"Hold on to my legs, Matti. I'm going to wash my head."

I stretch down from the stern and immerse my head in the Atlantic, feeling Matti's firm grip on my legs. I blow bubbles and massage my scalp under the surface. Suddenly this is all very humorous. I've been in a lot of scrapes in my lifetime, but never here, tired and hanging from the back of a boat with my head in the middle of the empty ocean on the way to Iceland. The distance to any offices, collective labor agreements, budget meetings, and sympathy strikes is light years.

"Oh man, that woke me up. I'm going forward for a little." I climb up to the foredeck along the starboard gunwale and rotate my arms, doing some calisthenics and just moving around. The sun is on the rise, and to the north-east is an orange haze. I start to come back to life, recovering my strength and my voice as my excitement increases.

"Damn, I was tired. Our route was starting to look like snake piss in the desert. We were getting nowhere."

"Yeah, it was winding, but I didn't feel like saying it every time so you wouldn't get even more nervy. It's that you can't change position at all; it really gets at you. How do you feel now?"

"Fine, really good. This helps a lot, and eating and everything. Let's get moving again. Forward ho!"

"Right on . . . hit it, Frank!"

I rejoice in the growling of the engine, in rising up into our hydroplane, in the rising sun, in the full barrels of fuel on the deck, in Matti's companionship, in Höfn somewhere in the distance, and in the motions of my arms, which are once again making the boat fly. My touch is light on the wheel and the throttle, allowing me to enjoy driving and beginning to check the log. I want the best possible speed. I squint at the decimal points on the knots. I am alive, joyful—I feel like humming, like singing.

"You want some sausage? I'll cut you some if you want."

"That would do me good, yeah."

"Here . . . oops . . . here you go."

"Thanks . . . Oh, yeah, that salty stuff's good."

Looking out for each other is important. These are important messages in our miniature world in the middle of the ocean. They tell us what they should and have nothing to do with the smarmy snobbery so common in the world.

The surge is coming on from the rear-starboard, and visibility sometimes drops to nothing. We plunge into walls of fog, seeing only a few dozen meters of waves as everything is covered in a watercolor grayness. Rain fronts pelt down on us. The height of the waves varies and the hull bangs away, but the *FinnFaster* charges powerfully north-west from wave to wave.

Matti is resting with his head against his seat with his eyes closed. He drowses and watches the instruments at the same time. He knows that he can't sleep and stays in control of his work. His thoughts are flying far away . . . systems . . . judges . . . legal proceedings . . . He has to remember it all. They're trying to get him, and there will still be many years of uncertainty yet . . . everyone was saved . . . and now we just drive.

It's good to be in the emptiness. The endless grayness visible around us surges, eternally in motion. It breathes, rising and falling steadily, coughing and hacking at times and then going back to its rhythm, as it has since the beginning of the world. The feeling of being free is dazzling—we are nowhere. There is no county, no village, no bend in the road, no street, and no postal code; our location is infinity, an intersection of longitude and latitude lines, two series of numbers on the screen of the GPS. Our location is not owned by any state authority. The emptiness was before us and will continue after us. We don't need anything else. We don't long for any shores. All is well.

The emptiness is safe. Here no one talks shit behind your back and no one digs pits for their neighbors. You aren't going to get a knife in the ribs, and no one whispers. I do not fear the sea; I respect it deeply. It has never threatened me. It wishes no evil, and danger is only ever created by my own mistakes. You cannot conquer the sea, and you cannot bribe the sea. With my companion, Matti, I am able to meet this majesty.

I drive and rest in the ceaseless banging and bumping—my exhaustion has disappeared. I accelerate and trim the bow, pushing forward and jumping, comparing how softly we come down at different rotations, hungrily striving forward and relishing how much distance is left to travel.

I consider the emptiness. I am very calm, in the middle of the North Atlantic in a small boat—I can't see any more than a couple of dozen meters ahead and don't know anything about the coming hours or day. I don't know about my future, not where I will live or how I will earn my bread. I only know that I am probably divorced and don't know where I will go when I return. Now I am driving, just driving. I know how to drive. A great unknown change is coming and this is where I am accepting it. Here I am strong. The bow railings rock and the rain and spray beat the windshield. The radar screen shines, empty. Why would I fear the unknown? Everything is possible, absolutely everything—even good things.

"Look right, Matti!"

A yellow dot pops into view to starboard. It disappears into the troughs of the waves for a long time and the rises up the slopes, high up to the crest, and then falls again.

"Let's get it. I want it. It looks like a buoy."

"We can't . . . what will you do with it, with an old buoy?"

"It's from here, which makes it important. Go forward, starboard side."

We rock for a long time in the three-meter swell; snagging it is difficult. Matti stretches over the side, and a couple of times he almost takes a bath, but his experience shows in his balance. In the end there is a bang on the deck.

"Your buoy, Sir." He grins.

"I'll take it to the cabin. Right next to the door . . . it's real, something I got myself."

At the same time we take our location and wolf down some chocolate. The entry in the logbook reads '09. 15 Fishing buoy 63°08,725'N 10°52,023'W.

"How are your hands doing? Don't you want some gloves?"

"The bandages help, and they don't ache anymore. I don't want gloves. I'm not cold, and it's more important for me to be able to drive better. See, these handles aren't just handles. The gas and trim are the finest fountain pens, and this is calligraphy, as I've always said . . . fountain pens and mittens don't go together."

"You do have quite the paws. I've been amazed at them for a long time."

Howdy

Our forward charge continues for hours and hours, but we do not tire. At three hour intervals we take our position, check our course, add oil to the reservoir, and add fuel from the deck to the number one drive tank. Everything is working, so we send our thanks with a storm bird to Harry in Bergen, one thousand kilometers back.

I listen to the engine as it grinds away, flawlessly growling at the stern. We think of it as a reliable friend struggling along with us, in humility. It becomes personified, as if it also has its own will and personality. Fondly, we call it Eetu.

We sing and talk over the thundering of the engine and the sea. A friend from years ago comes to mind.

"Listen to this."

"Yeah?"

"I was thinking about this one friend, a really great guy. He was hired by this one company as the CEO. Then a few years later they didn't like the look of his face anymore and he suddenly got the boot. Well, luckily he found another good job and continued there."

"That's good . . ."

"Keep listening. A few more years passed and then the same firm needed a chief financial officer—they were reorganizing. Guess who they asked to do it. This same guy. He came gladly, letting bygones be bygones."

"That's a strange twist."

"Well, let me go on. A few years later they let him go again, even though he was a top performer, because they didn't need him after all. April Fools! Think about it: Two times they hire him and two times they give him the boot—in, out, in, out . . . A professional, but he doesn't play games or plot and actually encourages his subordinates. I think the company probably botched it both times, either in hiring him or firing him. What do you think about that? How can anybody believe in anything? How can a guy pull himself together and find the strength to start over, especially if he's getting up there in years?"

"How did they allow that?"

"Allow what? They did it themselves. These asses call it 'dynamic business culture' when the staff changes faster than the products."

"Damn. Some people need their balls tarred and good dragging under the keel . . . Repeatedly . . ."

We continue in silence for an hour, our speed around seventeen knots. The fog has let up, and the waves are coming from the rear starboard.

"Hey, look. There's something ahead and to port. Is it on the radar?"

"Yeah, there is something there. . . ."

I steer towards it and the dot grows. We begin to see masts, and foredeck and stern structures come into focus above the swell.

We quickly overtake the red-hulled, fifty-meter fishing vessel. She is hauling something: Large balls plough through the water behind her, and her course is diagonally to starboard from us. My heart leaps—meeting other wanderers in the middle of this emptiness is a huge event.

I remember the con man.

"I'm going to hail her . . . I'll ask them in passing how the fishing is. . . ."

"No, don't hail them. They're working . . . well, let's just give them a good hello!"

I speed up to the limit the waves will allow, and we pass twenty knots as we sweep by a hundred meters off the starboard side with our hands up in greeting. The three men working on the deck drop their ropes and stare after us. They just stare and stare . . .

Thinking about the meeting from their perspective keeps us entertained for a couple of hours. First a dot appears on their powerful radar. Then a point appears on the horizon. Then a Finnish flag the size of half a sheet peeks through the waves. And then an aluminum motorboat speeds by the ship with two figures in red suits waving at them like we were just two pleasure boats passing in our home waters. Then just the blue and white flag plunging straight north-west into the fog . . . in the middle of the Atlantic.

We spend a lot of time guessing at what the crew must have thought, and the version that wins the final vote is the one in which the skipper rubs his eyes and says gruffly, "Break out the rum, boys. . . ."

The contrail of a jet is visible for a moment. No one in its cabin knows anything about the microscopic point struggling northwest, around the clock, ten kilometers below. Splashes of foam fly rhythmically to each side as it dives into the troughs and climbs proudly to the crests with its swirling wake boiling in the swell for a fleeting moment, changing the course of the waves not at all.

At the rear of the point there are two even smaller red points side by side.

Fulmarus glacialis

"Vatnajökull on the horizon. It has to be!"

All the trumpets of the world sound in my ears—tenderly yet intensely, and resounding with joy. I stop as we gaze on the heights of the eternal glacier. We rock in the strong, smooth, turquoise surge. I maneuver my way up to the foredeck. I lie down and lower my camera to the waterline. I'm trying to photograph a lone bird curving by out in front of the bow with its wingtips trailing on the surface of the water. It is a storm bird, *Fulmarus glacialis*. I learned that from Matti.

The sky has ripped open and the sea is calming quickly, the bright sun warming the surface. The deep turquoise color arrests my gaze over and over.

The boat rises and falls slowly. We are close, so close. Just a few hours left! Eetu is quiet as the waves lap at the sides.

I know that I can drive the rest of the way. The boat will hold up. The engine will hold up. We have enough fuel. We have enough energy. Matti is dozing again. I accelerate slowly to not wake him. I can do fine alone; Matti can sleep until we get to the Höfn Gap.

I glance to the left: A lined, weather-beaten face with an overgrown, unkempt beard and a wet beret over his eyes. I can hear a light snoring. He is tired, as am I.

Matti, I won't wake you up—let me just add to your dream that we will be pulling in soon. An hour later I'll wake you up. You have Heikki's directions for steering through the difficult strait. You know them by heart—I know. You are my navigator . . . in fifteen minutes I'll wake you up. Matti. You gave me every course. You did what you promised. I just got to drive, to drive and crank away at the controls . . . for so long . . .

"Matti, wake up! We're coming in. The Höfn Gap is right in front of us!"

Is It True?

Driving in the powerful wake is good. The back of the boat reads *Hornafjördur*. She is white, just under ten meters, with a flat stern, a strong bow, and a cabin up front. A sturdy targa rises above the cabin with lights, spots, and antennas. Two half-meter winch rolls are attached to the railings in the open space at the stern.

The boat's diesel rumbles powerfully, with its exhaust gas dissipating to port. The gap is approaching, and the current is strong, just as Heikki reported. Above the gap rises a dazzling white apparition, the icy top of Vatnajökull.

I feel impatient—we are finally arriving in Höfn, my spiritual final port for the last two years. Right now it is opening up before us beyond the gap. Our eyes take in the scene greedily.

A few dozen large trawlers rest at the dock before us next to the fish factory having disgorged their innards. Heavy and unmoving, they are waiting for something. To the left, the back of the bay is dominated by a low, white office building, probably the harbor bureau, since there is a flag on the roof. Behind and to the left in front of a gravel bank along a jetty are dozens of fast, powerful, orange and white fishing boats, just like the one we followed through the gap. The bow of every one is pointed outward.

They are grouped straight, back-to-back and side-by-side like an anxious horde ready to shoot out after their prey as soon as the message comes in.

I can see that life here is about fish—the harbor makes it clear that this is at its heart. A slight wistfulness takes me as I guide us slowly to the dock. My vague image of an ancient harbor with all its nostalgia is gone. People rush to the dock. Sport utility vehicles start up crankily and rush in sprays of gravel around the bay towards our dock. They have been waiting.

I turn off the engine and for a moment we can hear the quiet. It is a beautiful sound. The shouts of greeting and the screeching of brakes are still far away. We have one more moment for ourselves.

"So this is Höfn."

"Yep . . . this is Höfn."

The people who come laugh and make noise, as if they had been waiting for us for years. They are unceremonious and loud, but at the same time considerate and friendly. Many of them have strong, sturdy facial features and their hands are stout and broad. They are already making fast our ropes.

"Come eat, come eat. It will be our treat."

The invitation wakes us up and we stretch and climb onto the small pontoon dock. We are in a strange situation. Nearly twenty-four hours of plunging across the Atlantic has finally thrown us into the place we have been struggling side-by-side to reach for twenty-three days. Years ago I found the name Höfn on a school map and decided that it would be the obvious harbor, if . . .

We come down slowly, busying ourselves with the fenders, securing the already well-knotted ropes, checking our flags, and turning off our equipment—playing for time. We do not yell or dance or carry on. We just go about doing almost trivial things, but we know we are in Höfn. That knowledge slowly begins to purr like a contented feline in our minds. I feel a peaceful, dizzying satisfaction, and my voice is a little rough.

"So where were you supposed to be going, Matti . . . do you still remember . . . where did we say we were driving . . . ?"

"Hold on, I'm thinking . . . wait . . . I think it was Höfn . . . or something like that . . ."

I can hear a quaver in his voice and slap him on the shoulder.

"Goddammit, Matti! Let's go eat!"

"You talked me into it!"

The phone rings in the boat, and I hurry to answer it.

"Heikki, damn. You couldn't have had better timing! The lines are tied. In Höfn!"

At the moment our friend is sailing with his own Iceland crew from two years ago on the Baltic Sea out past Utö towards the Danish Straights. Hanski, Vesa, Pekka and Heikki share this once-in-a-lifetime moment with us there far away, and despite the distance they are with us and we with them.

Climbing out of our survival suits feels strange. Our suits are easy to move in—they don't pinch or rub, we don't sweat or get cold in them, and they have become our basic clothing. The harbors have just been exceptions. But jeans, a windbreaker, and deck shoes feel strangely exciting, like the suit I wore for confirmation thirty-three years ago.

"Of course you will drive here. All of Reykjavík is waiting for you, for the celebration . . ."

"We're watching the weather. It's getting difficult again. Even the fisherman are on land. We aren't playing any games."

"We have a surprise for you . . . in Finnish it goes . . . K-A-U-N-E-U-S-K-U-N-I-N-G-A-T-A-R! Beauty queen!"

The telephone crackles. I stare at the dock, gritting my teeth and pursing my lips. I barely manage to keep a grip on myself.

"I'm not promising anything. We'll make our own decision. Thank you in any case, but we'll be watching the weather."

Matti and I review the situation quickly. We aren't about to start clowning around for the marketing department, and certainly aren't going to be posing with beauty queens just because someone else wants us to. We are in simple, modest Höfn, our original goal, a barren place like Kökar, and that suits us. There aren't any glittering nightclubs here or celebrity chicks young enough to be our children. We aren't patsies. All the fuss about beauty queens is for people with empty heads. None of that has anything to do with humanity, femininity, or the real beautiful things in life. It is a brutal, shrill marketing lie, a chorus shrieking hollowly. No, we certainly are not going. There will be no negotiation between the two of us or anyone else.

The clear summer night is bright—we're as far north as Oulu in Finland, nearly to the Arctic Circle. The harbor is sleeping in the gentle light. I look up at the crest of the glacier shimmering in the sunlight, and I can still hear Muru's gushing:

"Is it true? Can it really be true . . . ? Oh, how lovely! Pekka, you made it!"

The dock creaks as I walk. The *FinnFaster* is resting in among the fishing boats. She doesn't look tired—her bow is ready for a new plunge into come what may. I think for only a moment and then quickly punch in the number. Elina answers calmly.

"It's me. We're in Höfn now."

"No kidding? It's so nice you called. I'm so happy for you, Pekka. Do you believe me?"

"Of course I believe you. It feels good."

"Of course it would. How did it go?"

"We went through everything, fog and rain. The swell was about three meters, and at first I was so tired . . . at the end we had a couple of hours of smooth water and got up above thirty knots, and the sun was shining . . . it was like a triumphal entry. How are the boys?"

"Oh, nothing new. Perttu is out driving again and Arttu is at football practice . . ."

"Listen, that call from the Faroes . . ."

"Let it be . . . we're just so different . . ."

"Yes, we are . . . of course we are . . . I mean . . . have you heard anything . . . about the papers?"

"Nothing yet. It will come when it comes."

"Yes, of course . . . well, I hope you're well there . . . at home . . ."

"You too. And give my regards to Matti."

I like sitting on dock pylons. I watch the tame eider drakes—these are the birds people throw bread to in Höfn. My thoughts fly from the ducks to Elina's delicious seabird casseroles, to home, and to our long relationship.

We understand one other and yet don't understand one other. Formal truces are possible, but they never last long, and I can't stand formal relationships like that. I need more—I'm slowly starting to realize that. I can't force myself to be anything other than what I am in the name of formality. It always leads to an explosion. The nausea builds up slowly and then the emptiness explodes in a burst of rage. It isn't good. I stroke her in my mind . . . so many years together though . . . we were young—we didn't understand . . . she was so flexible . . . we had so much fun together . . . she was so kind just now . . .

I sit in silence for a long time, listening to the wind. It grows in strength, and continues . . . everything continues . . . for a time . . . I have come to Höfn with Matti; this is where we said we would come . . . it's good to be myself . . . for the first time . . .

The Sigurvindur

Smoke from halibut being fried in butter spreads through the narrow cabin of the *Sigurvindur*. It rises from Torkel's skillet, circling around the lamp burning on the ceiling, licking at the edges of the four bunks and then falling to tickle my hungry senses provocatively.

Hallgrimur asks me something—I don't understand, and, laughing, he swings his hand in the air in a wide arc and bangs the top of the table with his fist. Then he claps me on the shoulders, looks me in the eyes, and then slaps his hand to mine again, his whole sturdy body exploding in laughter. His weathered face is grinning widely, every wrinkle exuding the joys and pressures of the life of a fisherman, the story of the foam and spray of the sea, weather reports, the imperative of the catch, finding the energy to go on each day and the trust of crewmates.

I am among my own. I look around: Bunks on the walls two-high, a table lower down in between, a hotplate and stove in the corner—a real fishing boat, cramped and homey. Smells of wood, salt water, the galley, work, and fish.

In the twenty-meter *Sigurvindur* they fish halibut; Torkel proudly showed me a 170-kilo fish in the hold, the largest they took yesterday. There was still a strong length of rope around the fork of the tail.

I was happy to comply with their invitation and am enjoying our time together. Fishermen are the same everywhere. Everywhere they look you in the eyes, state their business directly, respect their friends and the sea, and don't kowtow to contrivance. We share the joy of our arrival with them. Often we spread the map of Scandinavia out on the table, and they look at each other, talking rapidly and then laughing like thunderous volcanoes. I am proud of their invitation. It was meant for us, as seafarers, not as some role players or as paid representatives of anyone or anything, and I am proud that people like them invited us.

". . . the daughters . . . were they angry . . . do you know . . . Ægir and Rán . . . ?" Torkel asks seriously.

"Yes, I know. Yes, they were angry . . . and also sweet . . . they showed us every side of themselves, Rán's daughters did."

They talk noisily—Hallgrimur is puzzled and Torkel shakes him by the arm:

"He knows, Hallgrimur! He knows the daughters of Rán!" With smiles all around, we toast.

This simple meal of fish in this narrow berth beneath the chattering mast wires is more festive than any I ever attended or hosted in luxurious surroundings as a necktied personnel manager. Between us there are no hard-fought contracts that either party might forget or cancel at an opportune moment, no pleasantries when we have a mind to curse or any attempts to wheedle out a neighbor's thoughts.

Our celebration is genuine, because it is based on something shared we hold within us.

The evening continues. We chat in a pidgin of Scandinavian and English, and all the memories of my 48 years of life glide through my mind. I am a prince, living the high point of my life. I met them in the fragrant cabin of the *Sigurvindur* at the base of the Hornafjörður in south-east Iceland. I knew that they were here.

"Why did you make your long voyage?"

He asks it very seriously, his rough voice lowered. My answer is easy, a matter of course.

"I wanted to meet you and your leader. I had to come."

Hallgrimur looks me in the eyes for a long time, unflinching. We don't avoid each other's gaze, and a peaceful sun shines on his face.

It Won't Stop!

The sunrise on the mountains is breath-taking. The Atlantic, alive with whitecaps peeks through the blue mountains and the ginger lava fields are cast in orange by the morning light. The whiteness of the innumerable snowy peaks mix with the clouds in the lightening sky.

The black and white four-wheel-drive SUV slowly struggles up the slope along the narrow, winding road towing our aluminum boat on a trailer. The combination of truck and trailer moves slowly and on the up-hills stops for a breather while the driver downshifts. At the tops of the hills the aluminum flashes in the low-angle rays of the sun.

Veku handles the six-ton rig with practiced skill. He has already been driving for most of a day from Reykjavík. We lifted the boat onto the trailer in Höfn and are continuing towards Seyðisfjörður, a harbor in north-east Iceland from which the boat to Bergen will leave.

Veku's family is beginning their trip back home after an extended vacation in this distant land. Fifteen-year-old Jonna is sound asleep in the back seat in her mother Anne's arms, and Matti and Veku and I are talking quietly up front. The boat, the car, and both crews are together for the first time. Our talk is genial and it feels good to be together. We followed our own routes from Helsinki to

Höfn as agreed, met, and are now continuing on together for a time.

I see something below in the valley along the shore.

"Stop, Veku, there's a barrel down there!"

The barrel I see has been thrown up onto the wide, sandy beach by the ocean, and I run the two hundred meters over to it up to my ankles in sand. And there it is: The oak barrel half buried in the sand fascinates me. The sea brought it here from somewhere, sometime. It had a purpose, it has a story which I can only guess at, and this is why I find it enchanting.

I saw it myself; I found it on a bright, early morning on the deserted shores of Iceland. I clean it and carry the barrel up to the car by rolling it along. I am as happy as a little boy who just found a real treasure! Matti looks at me and laughs.

"What do you mean to do with an old barrel?"

I'm covered all over in dusty sand. While I try to clean myself off, I look at Matti in confusion. Can he be serious?

"I'm taking it along. This is amazing. Look at the oak boards. It's almost intact. . . . Where could it have come from? Where did the sea find it? This is incredible. Does it still smell like anything?"

"I'll be damned. It's always something with you . . . hehe . . . dragging all kinds of old junk around . . ."

"Throw it on the trailer! This is definitely going to Kökar."

Jonna is awake now and comes up to sit next to Veku. The four of us still fit, but only just. According to the map we are at an elevation of six hundred meters—the descent down to Seyðisfjörður waiting below is about five kilometers of steep, winding downhill.

We start down slowly, the sturdy bow of the *Finn-Faster* obscuring the rear window. I glance to the right, seeing only the sun and the gray mountainsides with their white peaks. The sun warms my face. I'm almost about to nod off and close my eyes. The sound of the engine disappears.

"Brake, Veku, brake!"

Anne's alarmed call jolts me awake. I glance out: Our speed has increased . . . what on earth . . . ? I turn to Veku and see him gripping the wheel, his face gray and his jaw muscles clenched.

He stares at the long straightaway ahead and splutters in a hoarse voice: "It won't . . . stop . . ."

Jonna screeches and clings to her father: "DADDYDADDYDADDYYY!"

The heavy vehicle and trailer continue to accelerate, plunging down the road ever faster, a grey wall speeding by to the side. The speed increases . . . seventy kilometers . . . eighty-five kilometers . . . ninety kilometers . . . We are in a nightmarish trap, to the right a ravine hundreds of meters deep, the left a jump of several meters into a lava field. The road is narrow and there is a boat behind us carrying eight hundred liters of gasoline.

"JUMP, PEKKA, JUMP NOW!!!"

I glance at Matti. He is staring intently at the road and trying to hurry me.

"GO ALREADY, NOW!"

I don't ask questions—I don't have time. I just trust him and open the door.

I look at the ravine speeding by. I can't jump; there would be no chance of survival. I see a 100-kilo man flying at high speed. When he meets the boulders he bounces and crashes uncontrollably and something breaks. It isn't a guess or a fear. It's just plain physics.

The rough asphalt flies under us—that's the only possibility. My right hand squeezes the door handle, my left the running-board, hanging in the air by my arms. No . . . ow, feet first as a landing gear, the soles of my feet on the road and my strong grip against the force of the road. I am a bomber trying to land with a full load on a damaged landing gear.

The road pushes against my feet, making me fight it . . . the only possibility . . . crumple, slowly down, legs crying out in pain . . . my shoes, new . . . sturdy . . . I concentrate, concentrate . . . the only possibility . . . slide, roll on the road, don't bounce . . . I'll let go before the bend . . . this is precise . . . a moment more . . . I'll let go soon . . . I'm ready, keep concentrating . . . down a little . . . soon . . .

"GO ALREADY. YOU WON'T MAKE IT IN TIME!"

I can feel the speed still increasing. I can smell smoke from the bottoms of my shoes . . . behind me the boat. I have to push off a little so it won't hit me in the back of the head . . . not too much with the ravine below . . .

I'm going . . .

My vision goes black and I curl up in a ball. The asphalt burns. I slide and roll for a long time, for an eternity . . . Finally I come to a stop and jump up. I am bloody and ragged. The truck and boat disappear, still accelerating with all the brakes smoking as it turns to the right.

In shock I watch as my friends careen towards destruction in a metal deathtrap. The mountainsides hear my bellowed cry of agony from the bottom of my soul:

"Help them!"

I glance at myself. Both palms and my right leg are leaking blood, and my clothing is in shreds. I try my teeth and find them all present and accounted for. Even my glasses are intact. My cap is on the road.

Good God, I have to get to them! I have to help!

I am out of breath and try to calm myself down . . . I can't run . . . if I run I'll end up more out of breath . . . I'll walk fast . . . I have to be able to act at the scene of the accident . . . maybe I can still help somehow . . . there will be a fire . . . for certain . . . the gasoline . . . a column of fire, an explosion . . . a heap of scrap at the bottom of the ravine . . . sirens . . .

I lope along taking the inside edge of each curve to save time. My mind is in agony and shock. If the Call of the Sagas is going to end in a tragic catastrophe, why was I alone left alive? I see the ocean, where we came from, navigating such a long way together, and now I am walking up here all alone. This can't be true. Veku was glad to come along. Why were this modest craftsman and his family destroyed in the mountains of Iceland? How will I ever be able to explain this to Niko, who stayed home?

Reaching each open stretch is agony. I try to prepare myself to see the shapeless wreckage below with its bloody bodies and explosive conflagration, to hear the cries and the sirens. I rush along in my unreal state, faster, faster. Black skid marks on the curves, thick and trailing closer and closer to the edge of the ravine.

Matti slams the door shut and drags Jonna away from her father, fighting at the wheel. He knows that nothing thumped under the wheels and glances at the speedometer: ninety-five already. Anne has fainted on the floor in the back.

Veku stares at the road rushing under the vehicle. The engine has stopped and won't start again—along with the brakes the power steering has also gone. Turning the steering wheel is like shifting lead. Because of the automatic transmission there is no gear braking and the hand brake is useless at this speed.

A corner is coming up, to the right. Veku pulls and pulls. He feels the trailer pushing from behind, and a cold sweat squeezes from the pores of his forehead. Matti has more space on the bench seat now and squeezes Jonna in his arms with his right side turned in the direction of travel.

The wheels howl as the truck comes close to the edge and the trailer pulls from behind, leaning heavily. The next straightaway increases the speed, and a sharper turn is coming up, the speedometer reading over a hundred now . . .

. . . this can't be happening . . . it's twisting so hard . . . the lava field, could I try? . . . the drop is so far . . . too far . . . the ravine . . . destruction . . . my family is here . . . Niko . . . at home alone . . . the curve is coming, impossible . . . just me at the wheel . . . so heavy . . . can't stay on the road . . . turn . . . turn . . .

"DRIVE, VEKU, YOU CAN DO IT! DON'T GIVE UP! YOU CAN DRIVE ALL THE WAY DOWN!!!"

Matti's roar reverberates in the cabin, and Veku squeezes the wheel and turns. All eight wheels scream out in pain, and the strange, heavy, leaning express train plunges down the narrow switchbacks of north-east Iceland. In the front seat two men stare at the rough asphalt speeding by and again at the next turn.

I've been walking for an hour with thousands of thoughts running through my head. I saw how they flew into the turn and am torn with agony. I can only walk, only prepare myself to meet the catastrophe. I consider options, how to act. I am alone in the mountains. How can I get them to the road? How can I get help if I make it in time? My phone is in the car . . . a fire would be the end. It will start immediately. I won't make it, but I have to try. . . . They were going too fast. . . . How did I make it? My shoes held up. . . . Where are they? Their pain . . .

The curves and angle of the hill ease up. I still don't see anything. I don't understand. I speed up. I haven't seen anything—my distressed gaze has studied every nook and cranny of the ravines and lava fields, and there haven't been any intersections, just the narrow serpentine road and the stripes of rubber on the curves. I begin to hope and lengthen my strides, my heart pounding. Stabbing pain in my hands and leg. I wipe the blood on the edge of my sweater. All I have is haste and a spark of hope.

The road bends to the left, and on the right a shallow river with gentle banks flows surrounded by greenery. I see the valley and the roofs of houses. There is something strange along the side of the road, something shiny. I freeze in place and squint.

"Oh, thank God . . . there they are, there they are . . .!"

The truck and trailer are still together, on the side of the road two hundred meters before the intersection in the village. Veku, Anne and Jonna are walking slowly towards me. I see it, but I can't believe it. Everything is in slow motion. I am rejoicing, but I also realize something and cry out in horror, "Where is Matti?"

I see him covered in blood, injured or dead somewhere. I grow frantic. I can't understand how they are walking or the boat standing there so tidily on the side of the road. Veku looks me in the eyes and, white as death, says, "Matti went to look for an ambulance. For you."

I Love that Woman

Matti plugs the *payazzo* gambling machine back into the wall. He was borrowing power for our mobile phone and calling home to talk to Pirkko. I see his flushed face. With his dirty windbreaker hanging open he looks like a vagabond. Matti's robust baritone reverberates through the whole waiting hall:

"I love that woman!"

The two dozen passengers at the Egilsstaðir airport jump. An old man takes his glasses off, raises his eyebrows, and looks at us for a long time. We are drunk. He isn't angry; actually, he smiles.

Our flight to Reykjavík is leaving in half an hour. Somehow I even managed to talk us into free seats.

"I love her so much!"

At home in Vihti, Pirkko had just heard about our morning downhill slalom run. Matti related it briefly. The tires were smoking, Veku was at the wheel with his teeth gritted, and Pekka disappeared from beside him onto the asphalt. The girl calmed down in Matti's squeezing embrace, and he urged Veku on:

"Drive, drive! You can do it! You have more in the game than I do!"

After an eternity of switchbacks, the heaving combination of truck and trailer came down into the valley and an hour later Pekka came steaming down covered in blood with his clothes torn to shreds. That was what Pirkko heard.

"We have an agreement. If something happens, the other continues on and lives a good life. We will be thankful for what we've had, and life will continue on. Always. Nothing upsets Pirkko, because we worked this all out when I started sailing."

Matti leans back in his plastic chair. The front of his wind breaker is gray with dried salt from the sea; he looks ragged, but I know his dynamite. He closes his eyes, remembering a short phone conversation from three years ago, before the sinking of the *Finnpolaris*.

"Pirkko, it's me, and I'm in a terrible rush. We're going to the bottom."

"Where are you?"

"West of Greenland. The ship will sink soon."

"Matti. You are my husband. I know you. You are the last person who is going to go to the bottom. Two weeks from now you're going to be here at home, like we agreed. You'll see. Get yourself up on deck. They need you there!"

Matti is snoring. Again.

Matkro

"Hey, that picture! Let's take the film to get developed! He promised clear as day!"

Matti just had a stroke of genius. The proprietor of our regular pub wants our picture and promised a meal for it.

My credit card has been canceled, and we're wandering downtown Reykjavík with our stomachs growling.

Örn Artnarsson opens the door smiling. The sign says "Matkro."

"Hello again. Your table is free. I'm holding it for you permanently. Do you want to eat?"

"Listen, we have that picture . . . here, do you remember . . . ?"

"Excellent. Wow, this is great . . . I'll frame it . . . there in the corner . . . now eat well. I recommend the lamb. Big portions for you . . . and beer, lots of beer . . . on the house. What a wonderful picture."

Our pub has atmosphere. It comes from the whitewashed walls of the small space, the rhythm of the sturdy, dark beams, the stout benches and tables in the booths, Örn's friendliness, and the generous froth of the brass tap behind the bar that gives wings to our post-mortems of the trip. Again and again we excitedly relive what we have experienced; our loud voices irritate no one—instead we are toasted and toast our friends in return.

"Listen, Örn . . . thank you . . . Hey, Örn, listen . . . that guy over there . . . he told us to do it again, since we just flew over the North Sea . . . just think, that's what he said . . . cheers, Matti . . . damn!"

"Yeah, you know . . . when somebody really knows how to drive . . . it feels great, such soft corrections . . . and then you just continued on . . . just flying . . ."

We're ashore. Curious, looking at passers-by, breathing in and sensing the friendliness of the people around us. Veku is on his way with the boat somewhere far away. We'll be driving again eventually. From Stockholm.

I look at Matti. His face is flushed as he looks for the right words, weighing, emphasizing, looking for words again feverishly, his arms trying to hurry his search for the declaration he is trying to make. He finds it, grabbing hold of it and proclaiming:

"That's it! We were so different that we were able to do it! Together. That's it! Only three men in the world could have done it."

"What do you mean?"

"I haven't met the third one yet . . . cheers, Pekka, skipper . . ."

"Listen, Matti. . . ."

I fumble for the words. This has been bottled up inside me for four weeks. I haven't been able to open the lid, not even just a crack, but still I have to get it out into the open. This thing is out there, and we both know it.

". . . well, it's just this one thing . . . how can I say it . . . when we were leaving, in Kökar . . ."

Matti perks up, looking at the edge of his mug and wrinkling his brow.

". . . I didn't trust you . . . I was under so much damn pressure. It felt like I was in hell . . . you were so strange, and then there was the technology . . ."

"I saw it. Those were you pressures. I didn't have anything like that to bear . . . Yes, I saw . . . it was tough . . . and that drive to Visby. Damn, I'll never forget that one!"

"Yeah, I was just so damn low and so tired. But now we're sitting here together and clearing the air. Look, when we were coming in to the Shetlands and I was fighting exhaustion again, I just thought, I'm going to trust. When I felt that I could trust you, it wasn't a decision, it was just clear."

"I saw that and I knew that you'd be driving again soon and get us out of that cauldron. I trusted too, always . . . Mykines . . . the way your hands moved . . . I've been watching them the whole time. You weren't even wearing gloves. How can you do it? We have to change that bandage before the meeting . . ."

"Cheers, Matti. Just think, I was so stupid I was thinking about changing navigators, like I was going to go on with someone else . . . that I might have left a guy like you . . ."

"Well, guess what crossed my mind. Should I hijack the boat and just go on my own?!"

"Hell no. Really? We were both out to lunch, thinking we could swap out partners. Damn we were idiots not knowing each other! I would have thrown overboard the best man for the job!"

We toast again. My bandaged hand hurts, but it doesn't matter. Our faded sweaters are lying on the benches. We speak the same language now, and it means something. Two thousand two hundred miles has opened us up. We talk. We listen. We understand. Our boat is far away, but we are still a crew—we have a bond. Bloodshot eyes, over the pounding ocean, nighttime arrivals at port, the endlessness of the fog and the gray emptiness lived side-by-side have opened our souls to each other. Below us is earth and around us a smoky pub, but our companionship continues.

"Tomorrow I'm going there and I'm going to get it."

Matti turns straight towards me and rubs his beard. His eyebrows narrow a touch.

"Don't get it, by God. When I was sixteen I was in Antwerp and ran out in the middle of it, just when he was starting. I just yelled, 'Keep your money, but you're not cutting me any more!'"

"I'm getting it anyway. I decided a long time ago. I decided that if I got here in an open motorboat then it would go right there, on my right arm. Then it would belong there. I'm going to get it because that's what I've decided and I'm not sixteen anymore."

John Pokk is waiting for me the next afternoon. He makes a small, simplified anchor on my right arm. I decided this a long time ago, and two thousand two hundred miles have sealed my decision. An anchor—hope . . . that I would never forget.

"You made quite a system. I've thought about it a lot . . . those sponsors and the radios and the message baton thing . . . the whole fix-up, damn . . . I saw in the spring how you were flying around making deals and organizing things . . ."

I remember the pressures of the spring, the negotiations and letters, all the naysayers. I also remember a brightly lit room in Töölö and the calm conversations during which so many strangling knots were loosened and so many oppressive mysteries solved. A grin sneaks onto my lips, and I let out a guffaw. I remember my thoughts again and it makes me laugh.

"What are you laughing about?"

"That sponsorship stuff. I got quite a few of them, but I forgot one possibility. It would have been pretty great. Imagine this: in big black letters on the side of the boat like on a hotdog stand, 'DR. TAPSA'S THERAPY CLINIC—YOU'LL GO FAR!'"

"Hahaha, goddamn it, Pekka . . . you're unbelievable . . . you'll go far . . . get them next time . . . cheers, Pekka . . . hehehe . . . hell yeah!"

Above the mirror light in the men's room is an electrical outlet. I plug the phone into the outlet and start punching numbers. I have to figure this out.

Full Faith and Credit

It's a sunny afternoon on the square in Ylä-Malmi, Helsinki. A second-story window opens onto a busy scene of stalls and café tents, people running errands, vacationers, and people at loose ends wandering around aimlessly.

The room has a restrained tidiness, and there are none of the usual trappings of those who relish their power. The lightweight furniture is bluish and the table is a light gray, with a few stacks of paper on it. A tall plant stands in the corner, and the door is open. A slim man of average height sits down in a chair and glances at his watch. A touch of gray at his temples shows that he has recently passed fifty.

Bank manager Raimo Hammar looks out. His next customer will be coming soon, an excavation contractor planning to upgrade his backhoe. He's a familiar face from over twenty years of acquaintance, a tenacious worker who started out with a single, old dump truck.

Hammar looks at the people in the square, thinking for a moment about his work and his area . . . thirty years . . . it's a lot . . . the last years have been hard; a lot has changed. The worst of the recession has started to pass, and now everyone is living carefully . . . trying to play it safe. His profession . . . a dirty word for a few years, the banks a veritable spittoon. It hurts sometimes . . . it's people's business we're trying to handle . . . it's important that we talk openly together . . . that we trust . . .

The phone rings on the table.

"OP Bank, Hammar speaking."

"Petri Carpen from Luottokunta Credit Card Servicing. We have a little bit of a strange situation with one of your customers, Pekka Piri. He just called from Iceland. Do you know him?"

"Very well. What seems to be the matter?"

Hammar's brow is furrowed. He has read that they made it, and if feels great. Pekka dropped in just before he left, to talk things over and make final arrangements, power of attorney and such. Hammar's brow furrows even more, and he stares out intensely at the vegetable stand in the square.

"He has a new card from us. Foreign charges have been coming in hard and fast every week, none of it being paid, so we had to cut it off. He's been on some boating trip to Iceland, and now he's calling from there saying that he had arranged with someone here that he wouldn't start paying until July. We don't have any agreement like that written down, so we had to cut him off. He gave your name; supposedly you know him."

"I do. Very well, his whole family. And have for many years. He stopped in before he left and we talked for a long time. I know the card, and he told me about the billing arrangement as well. Since we're speaking openly, I'll say that he's had some rough times, but he's always handled all his business properly, and I trust him completely."

"You mean you think we could open the card again?"

"Absolutely. That is my opinion. Agree on a sufficient limit and help those men get home. I have no doubt the payments will be paid in good order. The trip these guys have been on isn't the sort of thing anybody could do."

"No, that's for sure. This was good to hear. On the phone he was pretty on edge. I'll call him back and set things straight. It was good I caught you. Thanks for the help."

The contractor is waiting at the door. Hammar glances out one more time, remembering that meeting before their boat cast off . . . it was good we talked openly then . . . otherwise I wouldn't have known . . . it was good they called . . . So, the boys have made it and will be back here in Finland after Midsummer . . ."

"Ah, hello, hello. Come on in!"

A long stall stands in front of a bench on a walking street in the middle of the city. Icelandic sweaters, stocking caps, and wool shawls fill the tables. The sun is shining, the air is clear, and the mercury is at about ten degrees Celsius. Many of the locals are already familiar, and we greet each other and stop to chat.

I'm waiting for Matti; we have a date soon. A month side-by-side, seventy centimeters away from each other night and day has left us with a need for some space to ourselves now. We move about together and separately, and if one of us leaves our modest hotel room without a word, there is no insult in it.

I saunter along the street. I think about Carpen from the credit card company: He called a moment ago and everything is arranged. He worked fast, and I knew I was speaking with a gentleman. He also conveyed Hammar's greetings. I look at the mountains for a long time, silently. They trust . . . they trust me . . . that is what this has been, at its best . . . just like with us, the whole time. That shawl . . . so beautiful. I'll take it to Vivi. It's like lace, a lace like life—black, white, and many grays.

The Leader

She stands erect in the middle of the large room. I look her in the eye for a long time. We both do. I sense a smile, an artless, genuine, bright smile. There is nothing to disturb us, and no reason to be nervous. Not the bandage on my right hand or our faded clothing. The silence is solemn.

She moves forward, and I do as well, our gazes steady, smiling. We approach slowly . . . approaching . . . and now we meet . . .

We shake hands for a long time, not speaking.

"Those were the grips of fishermen and sailors, real, solid handshakes. I always recognize them."

I look at her. I am not thinking about my words but my business, so it is easy to speak.

". . . esteemed President Vigdis . . . Dreams are man's happiness . . . you see before you two happy men. We have met your people and now you, their leader. . . . We have come from afar, ourselves . . . I wrote long ago, and you answered. Your invitation helped us. . . ."

My gift, a black and silver Finnish *puukko* knife ends up on her desk. "That will be its place. It has such a unique voyage behind it," she says, emphatically but calmly.

Matti has a card, the one he was given by the old man on the dock in Kökar so far away. Matti tells the card's story. The president is moved, looking at the skillful woodwork and then quietly saying, "Yes, this is exactly how messages once traveled . . . very long ago."

She decides to write a personal thank-you, and her secretary assists. The letter is enclosed in an envelope bearing the presidential seal.

The four of us sit for a long time around the 500-year-old table in her office. She gladly accepts the Nordic message baton we have been carrying. We speak quietly as she asks us all about our trip.

"Your hand—how did you injure it?"

"Out there . . . I scratched it, somewhere."

I think of the newspaper men. It was good I didn't bring them along. There wouldn't be room for them in this moment. They would want to come snap away with their cameras and try to find out the exact time and place of our meeting. We didn't tell. We answered by saying, "We made it this far with just the two of us, and we'll make it the rest of the way just as well." We were profoundly grateful for our invitation and didn't want to sell it or sell her.

The low twilight of the evening filters through the arched windows. We chat about journeys, the sagas, history, and humanity. She is in no rush, so we are able to exchange thoughts at leisure. She looks us in the eyes, listening and respecting what we have to say. Through the window I see a small, white church whose age no one knows.

We feel welcome—she shows us that. The ten minutes we had been promised ahead of time has grown into an hour. She appears not to be in a rush to get anywhere. She confirms the human value of her visitors. It does not come from on high, but from right there next to us, naturally and with obvious consideration as she listens.

I realize that I am talking with a head of state in the dim light of evening, sitting at an ancient table. The sagas are true—the Call of the Sagas has been fulfilled. Two years ago a brown envelope appeared in the mail in Tapaninkylä, Finland. The return address read "President Vigdis, Iceland." I desired and dared and finally got to meet this leader for whom I have so much respect.

"Thank you for your visit. It was lovely to meet you. I will remember this."

I carefully glance at the previous page of the guest book. I see the signatures of all of the heads of state of the Nordic countries written there.

Do You See any Motorboat Down There?

The terminal at the Reykjavík-Keflavík International Airport is lively. Before us a trim, dark-haired woman in spiked heels clicks along pulling a wheeled bag. Behind the makeup and stylish clothing I can see that she is going somewhere—she knows her goal. I feel sympathy. But does she really know her goal? What are her own goals, and what have been given to her by others? What does she want herself? Is she completely sure, despite the outward shell?

Announcements echo through the hall, instructing passengers where to go and what to do. We are no longer at the controls of our own craft being battered by salt water—we are back in the system, and it feels strange. I look around curiously . . . is this really what it's like? I had forgotten. . . . We just have to listen to the instructions and walk along with the others, surrounded by imitation leather and chrome. . . .

It is very strange and very exciting, but lonely in a melancholy sort of way.

The blue and white Icelandair Boeing 737 has rolled to the end of the airstrip and the roar of the engine test fills the passenger cabin. I am at the controls of the *FinnFaster* again. Ahead of us is a long leg. Let's look over everything together. . . . Have we remembered everything? Fuel OK? Safety systems, GPS, gauge calibrations . . . Are you sure you remembered everything? Are you sure you're ready?

We ascend rapidly and the plane banks sharply to port. Lava fields and icy peaks peek through a break in the clouds—I feel sad leaving you. I am at the mercy of this external force, which is taking me away, so far away, back where I came from. Sagas, you will stay here, but you will always be with me. I found you; I came to you myself.

The roar levels off as we set course for Oslo. Somewhere below we pass the Faroes, Mykines, Tita and everyone else. The young people who gathered around to listen to Matti play the guitar are going about their business as we simply brush on by, high above. The system is carrying us away.

The clouds and fog thicken. I am both present and absent, warmed by the whiskey. Somewhere very far away the *Sigurvindur* is plying the waters once again. Perhaps Hallgrimur and Torkel will find the 300-kilo halibut they have been trying for. Perhaps . . . we are all getting older, but perhaps even so . . .

The foggy Atlantic is surging beneath us, living its timeless life. The fog, the currents, and the waves continue along their courses as they have for millennia. They grayness continues. It is far away, but very much present. I remember everything and always will. It is warm in the airplane as we just wait and doze, even though we aren't accustomed to doing so. We haven't been able to move an inch without struggling until now, but now we order whiskeys and move along thirty-five times faster, simply trusting in others. It's strange.

The jet engines thunder in the sky as the airplane stands in place and the clouds and Scandinavia and all the history of my life rush towards me. Technology is flinging us over the North Atlantic at one thousand kilometers per hour. The crossing now lasts three hours—the same journey that I needed three years to cross emotionally.

The businessmen in front of us are satisfied with themselves. They had a successful trip, made good deals . . . As they toast each other, I hear that they are Swedes, from Gothenburg. Their enthusiasm is considerable. They have succeeded. . . . I look at Matti. He is sleeping with his beret on his knee and a piece of fabric on the breast of his faded sweater. It has the blue cross of the Finnish flag and the words "Matti Pulli, navigator, m/y *FinnFaster*". Every furrow of your face, every hair of your beard is familiar and I know what you will say as soon as you wake up. You, Matti, were by my side for every mile, every wave. You did what you promised, struggling and enduring. You know. Don't allow our neighbors to bother you—they have their joys and we have ours, which nothing can ever take away . . .

Matti wakes up. Lost in thought, he looks through the clouds at the Atlantic swells ten kilometers below. He is quiet, just looking down into the grayness.

I nudge him.

"What do you see? Do you see any motorboat down there heading towards Iceland?"

Matti turns to me slowly and looks me in the eyes for a long time. I hear his gravely answer and see no waver in his gaze:

"No, and it will be a long time before I do. I probably never will."

Welcome Back

The storm rends the trees, making the dense boughs scream for mercy. The canopy over the cockpit of the *FinnFaster* explodes now and then as sheets of rain pelt against it. The Gälarbryggan docks on Djurgården Island in central Stockholm slosh in the evening darkness. The light of a candle gleams weakly through the window of the boat. I'm outside wandering around the dock anxiously with water streaming from my windbreaker, looking, but not seeing it anywhere . . . I set it right here, I'm sure of it. Right next to this gate, right here, two hours ago . . .

Disbelief slowly turns to distress and distress rolls on to rage as I clench my fists and swear and bellow from deep within me:

"Matti! The barrel has disappeared! Damn it all to hell! Someone took it!"

"It can't be . . . isn't it there?"

"No, it isn't anywhere. It's disappeared. I put it right here. Goddamn rat bastards!"

Matti stares through a gap in the tarpaulin at the empty dock. He knows the value of the barrel. He knows that I told Muru about it, as excited as a little boy, that I heard her bright ringing voice christen it the Barrel of Dreams found in a faraway land.

Matti stares at the gate, his eyes blazing. He takes a deep breath. The air rumbles with rage, and a rending bellow makes the metal net shudder:

"OUR BARREL! THIEVES, LEECHES, RATS!"

The barrel is gone, and all we can do is fume, powerless. I am inconsolable. Because there was so much connected to it. Because I didn't remember that I had returned. Far away we could trust. I didn't remember that here you can't turn your back without the jackals striking.

We pull our caps down over our ears and tramp off to the pub with a solemn vow that if we see the barrel, whoever is carrying it will get tossed in Lake Mälaren. Gnashing my teeth I decide that I won't give up. Somehow, someday, I'll get it back. I am tenacious—I'll think of something. The barrel belongs on Torrskata. It's mine, wherever it wanders.

The 26-meter-per-second northerly storm lets up in the morning to 16 meters-per-second. We decide to leave. Ahead of us is the surging Sea of Åland, and at our destination friends and the cabin await us. It is Midsummer Eve. I call the coastguard on Kökar.

"We're leaving in half an hour. We'll come in through Ören. The water is still rough—the old sea is swelling and the northerly is still blowing. We'll be there between eight and ten."

"Great. It's good you called, so we can know to look out for you. And, welcome back!"

The familiar voice from across the sea is startling. Suddenly Kökar has moved very close and became very real, making me anxious to get there.

The Sea of Åland is a trial, since driving is new again after our break of ten days. The swell of the Baltic Sea is strong and high, and the waves are irritatingly short and steep. We push our way forward, but I constantly have to reign in our speed, even though the gas lever always finds its way back up . . . Midsummer Eve, Kökar!

The sun is already glimmering from astern, and my shoulders are aching. Again we see no one; we have driven in solitude nearly our entire journey. My thoughts fly far away again to Torkel and Hallgrimur . . . Midsummer . . . driving on the Baltic . . . were we really there? Vigdis, the leader . . . we got to meet her . . . the twilight, the table . . .

The hull thumps and shudders, the bow pushes up high and then crashes down on the next slope, foam flying over us and Eetu growling grumpily.

"Matti, there's something there. Is it moving?"

"It's a ship, but it isn't moving. It isn't a cargo ship." The dot grows and familiar orange stripes appear on her hull.

"She's one of ours, Matti."

"Right you are."

The coastguard station ship *Tursas* is out alone guarding the border of Finnish territorial waters. It is imposing in its massive, silent gray austerity. The Finnish flag captures my gaze for a long time—I just look at it, because it is so my own. We are close.

The struggle continues as we feverishly devour the waves with Kökarören just a few hours away. We are waiting for it, watching, scanning the agitated horizon and checking the clock often . . . it is waiting there somewhere . . . as a friend, as a sign.

"I see it! Look at those high splashes! There it is! Ören, Ören!"

Timeless, ageless music plays in the chambers of my mind. They came through here long ago with their rattling and shields and sails, and this was the place where we silently set course for Visby. While I have been gone the waves have polished my islet into a solemn monument.

"I'm going to head straight in along the western side. I can do it on my own. Just watch for any logs left from that Estonian barge accident so Eetu's prop doesn't get smashed. I'm going to get up on the crests, haha, Matti, we're going in . . . Karlskär . . . There's Jusskär . . . the radar tower . . . *Oh, when we made it to the dock that day, we carried dad to melt that day* . . . haha-haa . . . *o-ho-ho, a wreetched waandering seeaman . . .*"

We're futzing around with the flags—the bundle is tangled. The harbor is waiting a mile away, hidden on the other side of the strait.

"... *many a land he'd seen that year* ... where is that C flag ... it isn't over there ... *the East and the South and the West he'd seen* ... where is it?"

"What do we care about one C? ... Look, that's Abersö over there. The first time I was here I got lost back there. ..... *brothers we're not though we work side by side* ... let's just go. The guest flags are enough ...

"*lalalallalaa* ... yeah, who cares about one C ... hoist the guest flags. We'll start with that. Take this ... Sweden, Denmark, Norway, Shetland ... ok ... Faroe Islands ... and then ... Iceland ... Good. Very pretty ... *Singing we sail the seas* ... We're ready. Let 'er rip, Frank!"

The familiar dock comes into view, two figures standing side-by-side looking towards us. The flags crackle as we sweep in at speed.

We lick the final spar buoys, and I drop our speed. The *FinnFaster* raises her bow, an eddy forming for a moment before the bottom, and then the hull settles down calmly. Our wake surges past us and we glide up to the dock. Everyone is celebrating Midsummer, including some loudmouth in a luxury boat who seems to think he owns half of Europe.

Vivi and Hemming are calmly waiting for us. As island natives, they were able to calculate our arrival time and aren't making any noise. Hemming grabs the rope nimbly.

"Welcome back!"

Vivi is sparkling, laughing, and hugging, and squeezing us both. Hemming stands next to us, grinning modestly, his eyes very bright and his gaze open.

"I picked some for both of you. They just opened . . . wait . . . there . . ."

Vivi sets fragrant sprays of lilac in the breast pockets of our faded survival suits. I look her in the eyes but can't squeeze out a single word.

"Can you wake him up? What if he's already asleep?" We are striding along a dirt road to a small cottage, Hemming moving briskly ahead with Matti and I following. Hemming's purposefulness is familiar. He is resolute.

"Of course we can with this kind of news."

Hemming knocks. The house is quiet. He knocks again, louder. A nightingale trills in the bushes.

We sit down on the old sofa. The old man's eyes shine.

"You have returned . . . unbelievable!"

Matti carefully digs in the leg pocket of his survival suit and clears his throat.

"We have a message. For you, Nils Volmar Sundberg. It comes from far away."

Nils Volmar guesses what it is. The old man is spinning like a top, wanting to start cleaning and baking us *pulla* cardamom rolls and anything else good he could do, practically bursting with joy. The envelope with the presidential seal lies unopened next to the vase on the table. He approaches it over and over, touching the clean envelope and talking and laughing.

The cork of the whiskey bottle creaks and we all say *skool*. Nils Volmar opens the envelope very slowly. His hands are shaking. He reads through the message again and again, stroking the paper and adjusting his glasses. He reads it eagerly again and the corners of his eyes grow damp.

"You took it . . . all the way! She got it! She got it! She thanked me . . ."

Hemming, Matti, and I look at the old man, stock-still, without stirring. A clock just stopped. In the summer night we see a happy old man.

The veranda of the cabin is bathed in a gentle early morning mist. I watch the eastern side of Tärnskär in silence. The brisk scent of wood—I have arrived, again. A cuckoo calls on top of the hill. Is that the place from which we headed south, setting off in silence towards Gotland, through those familiar rocky islets? Was it me who came here to my cabin as I was leaving, opening the lock and hastily scribbling a few lines under pressure in my log book before leaving, just to be sure? Was it me?

The stove is blazing—I am too excited to sleep, having come from so far. Sleep, rest my dears, all in your homes. I can stay up, I want to stay up, so you all rest . . .

The call of an eider in the bay, the sun reddening the edge of Tärnskär, and the morning mist hanging over the juniper bushes. I look at it for a long time. The smooth rock faces on the hill begin to steam.

Old Mustachioed Boys

Hanko Peninsula falls behind to port, and the silhouette of Russarö Island blurs as we push on towards Porkkala Peninsula. Approaching the last goal of our journey is exciting and exhausting. The Estonia Basin is a different kind of harbor, our inescapable final destination, only hours away now.

The Gulf of Finland grants no pardons. I am struggling at the extreme limits of exhaustion as we rock and bump along, spray flying over us as we grope our way forwards. I am driving at the limits of what the conditions allow, finding myself pushing the gas forward eagerly every time I slow down after another hard hit on an oncoming slope. We are silent, just staring as the horizon bobs up and down between the bow rails.

I don't know where I'm returning to. The North Sea, Mykines, the Atlantic, and Höfn fill my mind, an intense joy suffusing every cell of my body, but in some moments I see only emptiness. Flashes of normal life.

. . . a home to be sold, maybe already gone . . . I gave power of attorney . . . no family anymore . . . I have to go somewhere, but I don't know where . . . bills—how will I pay them— . . . I've never gotten divorced before . . .

The struggle continues, the final hours on an angry sea feeling harder than the ocean stages that stretched hundreds of miles. Now our craving to reach this final port lends a mad urgency to our forward plunge. I become anxious . . . no more errors . . . still have to find the energy, just a little more . . . stay in control. . . . hands . . . drive properly . . . to the end . . .

We crash violently into an oncoming slope, and I draw a deep breath. We rock in the strong swells, waves buffeting the boat and carrying her along, sending water splashing over us. Far off, Porkkala is dimly visible as a narrow, bluish strip rising above the water.

"Let's head in, Matti! This just keeps on coming. Take Söderskär—let's get inland already!"

"We don't have that map. With the coast we can't . . ."

"I know it by heart. Just get me in there. These are my waters."

We curve in along the west side of Söderskär Island, one of many that bear that name. I look at the familiar cove in the evening sun, and a distant splash of euphoria fills my mind.

"Look Matti, over there, to the right . . . We took the family there . . . so many times. Over there is where we pitched the tent. We would swim, and once the boys and I whittled a sailboat . . . they gave it such a crazy name, *The Ginormous* . . . we sent it out into the sea . . . over there we dropped some nets once . . ."

I am happy and excited, replaying my family's favorite memories, having them surround me tenderly. Out on adventures again with my hedgehog-haired sons, looking at insects do their mating dances over still waters with Elina by my side, bringing a salmon back to my family on a clear, late autumn night.

We float for a moment in the calm water near the shipping lane. The sun warms us for a moment more. The engine is stopped. We enjoy a short break and prepare for the final swoop in. The canopy is down.

We just look on in silence, enjoying the view. The crashing and banging have passed. A sailboat glides in to Lähteelä Island.

"Hey, Matti. Something just occurred to me. We were out at the right time."

"Indeed we were."

"We're forty-eight. I think that at thirty-eight we wouldn't have been up to it yet. And at fifty-eight we wouldn't have been up to it anymore. It seems that way to me anyway."

"There's something to that. . . . It was a good time. Yeah, I think you're right."

The *FinnFaster* cuts aggressively along the smooth waterway. Water hisses on the sides and small irregularities in the surface patter on the bottom of the hull. Eetu's muffled growl is oozing with confidence at his victory, and the flags above flap out a joyful drumbeat. We stand side-by-side, exploding with laughter and waving to everyone who passes. We get up to thirty-two knots, but I'm still keeping a few in reserve.

"Hahahaa, the Neste Tower . . . goddamn . . . heyyy, Matti! Ryssäkari ahead, heyeyey Ryssäri! . . . Ayayay . . . *ooold mustaaaachioed booooys are we . . .*"

"*. . . both meeen and seeeals beeear them weeell . . . oh!* Pekka, come in by Kustaanmiekka . . ."

"Absolutely, right by the fortress . . . Here we go, Matti!"

I glance at my companion. This middle-aged sea captain is clowning around with me like a little boy. Cracking wise and telling dirty jokes and yammering away, clapping each other on the shoulders as we laugh, but our hands have the strength and firmness of grown men. They know what they are doing—they aren't fooling around and keep us on course.

My wave washes over me again. It is a powerful mental tsunami: I was there. I was really there. We drove to Höfn! Muru . . . laughed . . . the day after tomorrow I will be a free man, the freest in the world. The Call of the Sagas was true!

At Kustaanmiekka I slow down considerably and sweep into Tykistölahti Bay, between the two main islands of Suomenlinna Fortress. We want to wait for a moment. We can't just go straight in. We let the boat settle down as our feelings tumble over us. We need a moment.

I dig my mobile phone out and call the voicemail number at the radio station. I dictate slowly, my voice rough from being so tired. A group of tourists walks over

Tykistölahti Bridge to the fortifications, and a trawler heads out to sea.

"... a brief moment ... that is all a human is ... so much behind ... soon we will dock, with new phases of life waiting ahead ..."

I still don't start the engine. I smoke a cigarette and drink in the warmth and smell of the land. I wipe the condensation off the windshield and dig out the handheld radio. I select the channel and press the transmit button.

"Ville Number 2, *FinnFaster*.

"*FinnFaster*, Ville Number 2 here. Let's head into the canal ..."

I remembered the police boat, the *FinnFaster*'s older sister. She escorted us out way back then after the trumpet playing and speeches. Why not ...?

The group of people standing around on the dock of the Estonia Basin see a police boat with a blue cabin sweep powerfully out from behind Valkosaari Island towards the shore at Etäläranta. In its wake a dull oxidized motorboat follows closely with flags fluttering and two red figures looking toward the harbor. The bows of both boats flash in the setting sun. The convoy comes in a straight line, right at the dock.

The Iron Ring

We glide in towards the opening at an idle. About ten people stand behind on the stone pier. None of them moves or speaks; each one just looks at us. Noises of traffic come from somewhere, brakes squealing, a streetcar clattering. I quickly look through the group, but I can't pick out individuals.

We are in the opening, and I disengage the propeller for a moment.

"This is the Estonia Basin, Matti, our final mooring."

"Indeed it is."

We speak with voices lowered.

"Matti, listen. We're going to be docked in a second. It feels strange. . . . Listen, before we go . . . thank you . . . for sticking with me, for everything . . . the whole trip . . . just the two of us . . . you know . . . and I know—we know—let's keep in touch."

"Thank you . . . yes, we know . . . of course we will stay in touch . . . always . . ."

The iron ring approaches; Matti is at the bow, ready with the rope. A little more. Just ten more meters. I look through the people, but I don't see Elina. So this was the moment I was waiting for.

The ring is right there now. I can see the nicks and rust on it, but it is still strong. Five weeks ago we cast ourselves off from it. Only five weeks? A whole lifetime!

Farther back I see Perttu's face. His sturdy, open young man's face is shining. Our gazes meet, and I can see him grinning at me. I respond, happy.

We are in. I look at Matti's hands. He lifts the ring and ties the rope one more time. The movements of his hands are quick and practiced. The engine clicks as it cools down, and I leave the cockpit and look at the dock.

I look for a moment more and then take a broad step off. I have the courage to face this now. In this step I feel joy and strength.

Postscript

The shovel scrapes on the gravel. I can see his back down below. A burly figure approaching his sixtieth year, in blue overalls and a faded cap, digs, throws a load up, and wipes the sweat from his brow. A blast of cold January air is coming from the north, straight over the Kihti Sea to the little churchyard on Kökar, and chilling us to the bone.

He digs again and then pauses to bend down, picking up another skull in his mitten and turning it in his hand, inspecting it.

"Pekka. I think this is Grandfather."

I know how artless Hemming is, so I believe that he did recognize his grandfather—he wasn't making it up.

Nestor's grave is ready now so we take the Heinekens out of the plastic bag. Grave beers—it's a tradition here.

"How many of them did we find?"

I count the skulls we dug up: Eight all together in the black garbage sack.

I sit up late, alone in my cabin. I can't sleep. I feel anguished. It isn't because of Nestor's death, the skulls, or contemplating my own death. I'm thinking of my sons. Those living examples of youth will also become skulls in the mud someday and maybe even be collected in a garbage sack. Me, all of us, everyone, even my beloved sons. This thought is the cause of my distress, leaving me wallowing in my own sweat.

The grave has been filled in, and we are together for a moment—Vivi, Hemming, Harry, Holger, Tage and I. The sun is dazzling above and the snow below. Nestor has made it to the harbor; his sailing days are through.

The contrail of a jet gleams in the sparkling winter sky. I stand watching it. I feel an enormous relief, feeling clearly that the skulls and garbage sacks do not have any significance. Life can only be lived today. Today is the time! That is what I want to tell my sons.

The Call of the Sagas was a part of my life and Matti Pulli's life. It was in our stars, and we followed it. To each his own.

PICTURES

. . . The dock recedes. Another look and a wave.

". . . First how about let's go to Harmaja to change the prop."

"a motorboat . . . and a ship—each one has its strengths, but the setting is the same, the sea is the same . . ."

..I feel a little absent. That isn't any good at sea. An oppressive feeling washes over me momentarily.

Lasse from Kökar's smoked fish for dinner.

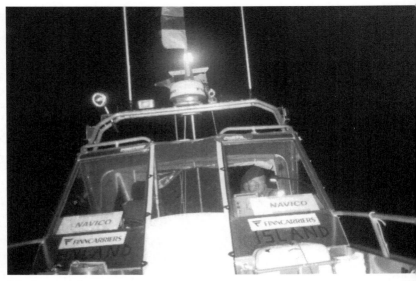

Night traveler

Menu 40 kilo salami for 40 days.

We raise the bow canopy; all we can do is settle down to waiting.

Another leg behind us.

Early morning in Kristiansand, Norway.

"The sun is sparkling from the left and splashes break behind us".

..this sea is alive in countless different ways, but it is rarely easy.

"They rush towrads us, slowly like the ringing of a church bell".

... Dawn breaks ... difficult seas continue, but we feel like we have th upper hand ... nothing can bother us.

Bergen, the pearl between the mountains of the Viking tales.

We position two barrels in the bow.

. . . FinnFaster is now fitted for ocean going.

"*. . . Right on. Let's hea[d]
over to Britain . . . and sa[y]
hi to the lords and ladies . .*

"... the master composer ... I pay homage to you ... this amazing journey of ours ... a solemn moment in the midst of it all ... we will continue ..."

"... Well, the mess-boy flew out onto the street ... and hard, goddammit! And, man, did that Scott yell! He was big ... ugly as sin ... goddamn!"

"...Rán's daughters tried the men, inviting them to their embrace ..."

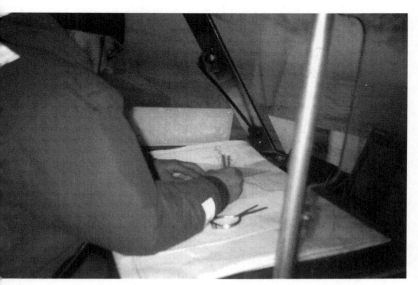

"... from the first waypoint there, that makes ... ok ... turn about ... and then ... next ... about there...."

We'll drain a barrel down into the drive tank using gravity, and then, later, the other.

The Middle Ages, traditions, the reserved Brits.

The smallest in the harbor in the Lerwick fog.

We can see the Faeroes, and the weather will surely hold, so we enjoy this time. This is the blue hour, with everything in balance.

... We drive straight into the colorful heart of Tórshavn ... again a new dock, again we breathe in a new environment ...

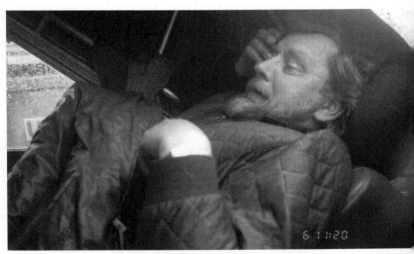

... The only important things are the proximity of a solid pier, the warmth of my sleeping bag, and the support of my reclined seat. The Frenchman's weather fax flits by at the edge of consciousness.

. . . he tries to roll over, but the seat is not a bed . . .

*. . That night we start on our final leg . . . nothing is certain . . . we will
 e alone out there . . .*

... we're swimming deep, but the engine is running ... The extra large drain ports in the rear bolt

Mykines is boiling.

...Muru, where are you... Elina, the boys... hard... death.... I'm struggling against him... goddammit... we won't leave... not without a fight...

Meeting other wanderers in the middle of this emptiness is a huge event.

... It is a storm bird, Fulmarus glacialis. ...

... the gap is approaching, and the current is strong ... above the gap rise a dazzling white apparition ...

So this is Höfn" "Yep . . . this is Höfn".

... *Climbing out of our survival suits feels strange ... The harbors have just been exceptions....*

"... *Why did you make your long journey?*"

. . Torkel proudly showed me a 170-kilo fish in the hold, the largest they took yesterday.

. . . the whole package on the way back will weigh six tons. . . plenty of opportunity to put the pedal down."

We toast again. . . . We speak the same language . . . . Two thousand two hundred miles has opened us up. . . . We are still a crew—we have a bond

". . . esteemed President Vigdis . . . Dreams are man's happiness . . . You see before you two happy men . . ."

YEARS LATER

I have heard the voice of the storm, I have heard the moans of the steel

My fate is bound to these, and fear of sinking I shall not feel

Bridge of the m/s Finnpolaris, August 11, 1991

MATTI PULLI
B. 1946 D. 2011

When we met at the herring market, I caught the bug for Pekka's dream immediately. I was taken with the intellectual and physical scale of it—it was challenging and tempting. For once I was seeing real daring. I've met people with all sorts of "intentions" in this world, but Pekka's calm, analytical approach and workable strategy were exhilarating, and his previous tour around Nordkalotten made me even more sure.

For a while I had been in an extremely frustrated place with my profession. The work felt too easy—it was a familiar routine—and the legal proceedings in Canada after the wreck of the *Finnpolaris* was a never-ending process of cynical quibbling. The torment of endless cross-examinations had the taste of cold systems and a million-dollar game in which they never ask you what you thought of first, the safety of your crew or your career. Seven years rolled by.

My mind was crying out for distance, for emptiness and space, for a sufficiently extreme emotional and physical exertion, by my own measures. I was literally ready to pull my boots on and walk to Kamchatka, the far side of Siberia, alone—it was that bad.

I committed to Pekka's proposition immediately—without reservation—because I could see that it was simultaneously crazy, realistic, and enormously enticing. It was downright tantalizing. I got my Kamchatka, many times over.

Our differences created a sturdy foundation for this difficult endeavor. They kept us active, alert, and curious. As professionals we both handled our work independently; the division of labor was clear and straightforward, and we both recognized each other's skill. During breaks Pekka left me in peace to do my calculations without asking any questions. A heading was enough. Myself, I enjoyed his driving—he was always in control.

Our differences provided additional perspective on solutions and stories, and in that sense they were a strength and an asset. Together we saw more. Our common commitment to a distant goal—sticking to our decision—was the same for both of us. Our professionalism and mutual respect strengthened the team and that team spirit increased our performance. This was especially apparent on the most exhausting legs of the journey.

The visit to Edvard Grieg's tomb in Bergen was a solemn moment. While our boat engine was receiving technical care, we were receiving spiritual nourishment. We stood at the feet of a giant. After than moment we were stronger and ready to head into the ocean stages. I will never forget that moment.

I think with great warmth of the 2,200 nautical miles we pushed through side-by-side. The legs driven together, the problems solved, the common trust, and the joy, ambition, grim determination, and opening up of your lives in the middle of the boundless expanse of the sea are still vivid and real in my mind after all these years.

In navigation terms, I consider The Call of the Sagas a very controlled voyage. In variable and demanding conditions, our work was logical, realistic, and emphatically professional. Our choices and corrections focused on making it to our goal.

Reaching Höfn, Iceland with Pekka Piri on the open motorboat m/y *FinnFaster* is the highlight of my 43-year professional career. And more than that: We became lifelong brothers.

Jokikunta, Vihti, Finland, October 22, 2010

The road ends at the sea, and there begins the high road

The length of the journey is determined by the distance of your own horizon

Estonia Basin, Helsinki May 17, 1994

PEKKA PIRI B. 1946

The relief and peace of the dock at Höfn made our return to land an overflowing of joy. We continued driving, reaching our harbors, plunging into the fog and mountain switchbacks speeding under us. We enjoyed our feasts with a happy lack of ceremony, repeatedly, calling and meeting frequently.

The changes in everyday life melted into the joy of relief—life continued after all. Muru was close for years. I had found laughter again.

The outbursts leveled off, changing to the relaxed smiles of the veteran, although now and then there was an uproarious grin as the memories flooded over me. The taste of the Call of the Sagas is still on my tongue. The aroma is still pungent, the colors are still saturated, and it reminds me of that place where I celebrated life: The surging gray of eternity out there in the empty vastness with a trustworthy and trusting companion at my side. He was dependent on my energy and my skill, and I on his. The memories are strong. Thankfully I have Matti. He knows—we know.

The anchor—a hope . . . that I would never forget. There it is, on my right arm, hidden by my sleeve, a simple, clean image. "A souvenir from Iceland," I sometimes answer when a woman asks.

The sea sighs quietly . . . Over the years, after our wake passed, sea radios have crackled with storm warnings, the turbulent currents off Mykines in the Faroe Islands have changed direction under a northwester at low tide, and the outermost skerry of Kökarören has continued to be polished down. The *Finnpolaris* may have settled into her final position on the bottom of the ocean, Matti's uniform coat may be floating fanned out in the captain's cabin as a nest for gastropods and nixies half a kilometer down. The *FinnFaster* is still at work cutting the waves, only with more oxidation on her sides than many others. She is no longer young—but does she still sometimes long for her sisters, the daughters of Rán?

The sea has lived from the beginning of time and will live until long after us—sighing and sometimes coughing; thousands of tales hang in its mists, including our Call of the Sagas. Perhaps some wanderer will hear a trace of the growling of our engine on the wind, or the singing of a strange yoke of men coming from the emptiness—someday.

The sea cannot be beaten; the wanderer is not a hero. There you journey with wisdom or without, under lucky stars or not.

The disappearances of many will remain a mystery, and every departure is a new story. The sea remains.

Pitäjänmäki, Helsinki, Oct 29, 2010

APPENDIXES

MARCH — WEEK 11

m/y FINNFASTER
H6/ Lars-olof Streng
847217 12531 + Matsin villaposta

14 MONDAY / Matilda

- Eemalo/Salo
- 8 Sauvuto
- + OMC Johansson
- + Herfasuma
- Beneton
- 10 - pachat
- +Tomi
- Westerlat
- H6 V12531
- + Rauean
- Pirkko Stenl
- +13 Horikon
- Tony Malmstr.
- - Böttny +
- - SATUIL
- 15
- 16
- 17
- 18
- .30
- Talogut.
- GR
- 4N

15 TUESDAY / Risto

- 8 Montten
- 9
- Peter Strandon
- +Alko
- Poike 52
- 11
- + Rita
- 13 Sosta
- R.76.486855
- Jari Sousston
- Rossi Jb Laver
- 2927530
- 15
- 16
- 17
- 18

16 WEDNESDAY / Ilkka

- 8
- + Ansgar
- 9 Koebo
- - Viking
- +SVR
- + Id. Hr
- + Kirje
- - Sosto
- 12 Sikatodi
- - Losk n R osh
- 13 Norden
- 14
- 15
- 16
- 17
- 18 Rilosk
- new todt
- - Ansi/tura
- + Marionte
- 21 mon
- 1358192
- Rita

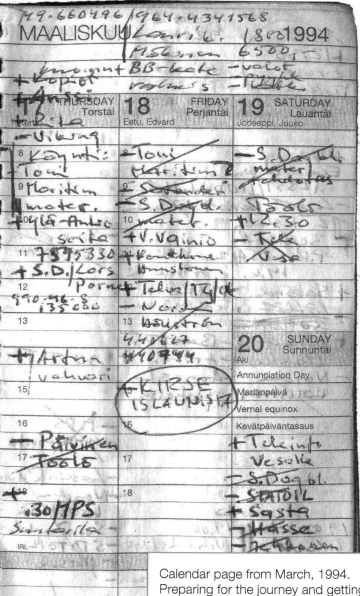

Calendar page from March, 1994. Preparing for the journey and getting the boat outfitted was a very busy time. Over six months there were five hundred negotiations. Everything focused on survival.

POHJOLA-NORDEN
Suomen pohjoismaisen yhteistyön keskusliitto
Centralförbund i Finland för nordiskt samarbete

17.5.1994

CERTIFIKAT

Härmed intygas att m/y FinnFasters besättning bestående av skepparen Pekka Piri och navigatören Matti Pulli är vårt förbunds sändebud på sin resa från Helsingfors till Island. Vår organisation, centralförbund för nordiskt samarbete i Finland, vill på alla områden stärka det nordiska samarbetet.

Pekka Piri och Matti Pulli har på sin färd med sig vår budkavle för att överräckas till Islands president Vigdis Finnbogadóttir, som uttryckt önskemålet att få träffa FinnFasters besättning efter ankomsten till Island.

CERTIFICATE

Hereby we certify that the crew of m/y FinnFaster, i.e. the skipper Pekka Piri and the navigator Matti Pulli, are ambassadors for our organisation on their voyage from Helsinki to Iceland. The Pohjola-Norden organisation, the central board for Nordic cooperation in Finland, promotes all forms of cooperation between the Nordic countries.

Pekka Piri and Matti Pulli brings with them a message which after the arrival to Iceland will be given to the president of Iceland Vigdis Finnbogadóttir, who has expressed her readiness to meet the crew of m/y FinnFaster.

Gustav af Hällström
Managing director of the Pohjola-Norden association

Pankki/Bank
SYP/FBF 200138-572306

Postisiirto/Postgiro
800014-71495

> On their journey, the crew of the m/y FinnFaster acted as ambassadors of Nordic friendship for the Pohjola-Norden Association. This "diplomatic passport" was an indication of this important task.

Torstarda

Hours	Region	Gyro	Gyro deviation	Magn. compass course	Deviation	Set and drift	Variation course	Variation	True course	Log	Distance	Weather
				Sumeles mile								
				"								
				8.6 – 74 WEDNESDAY								
			SÄÄ TAVALLE	Suuret mile	Myrskyisä		NW					
			VIRRAT AIKEUNAWEET		KOVIN		VAIKEUHEIR					

Noon D.R.

				= RAJU KOITOS =								

Tank soundings — **Sounding of bilges**

Navigation lights: On hours _____   Off hours _____

TUESDAY The **7th** day of **JUNE** 19**94**

TÖFT

7055 Lóurii 1668'   I-Toshihi, 162.5

1115 P= 62° 04.5' N ∧ 7° 40.4 1686.0
6.15 D= 236'
      RANTASOLTU TAKASIO TORS HAVEUU

Lat. −1732.9'—   Long. 1732.9 / 13
                       1660
                       72.5' nm

                       9415'
                       ~ 79 km

4 Liter össg —
3.5

m/y FinnFaster log book for June 8, 1994, after the night near Mykines. No ship has every returned from there before during low tide with a northwester.

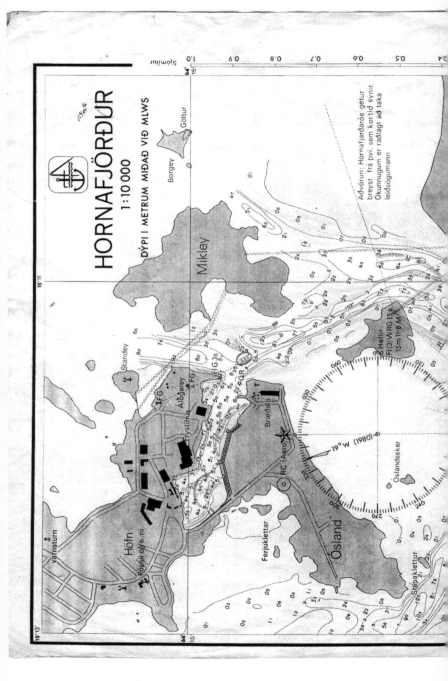

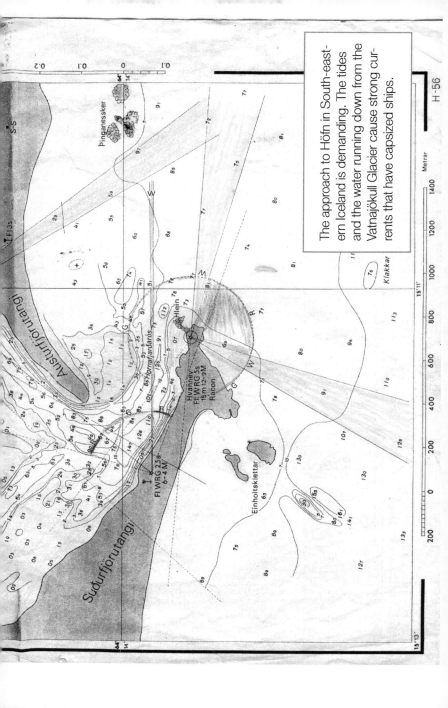

The approach to Höfn in South-eastern Iceland is demanding. The tides and the water running down from the Vatnajökull Glacier cause strong currents that have capsized ships.

Pekka og Matti Julla skipstjóri á Höfn. Á innfelldu myndinni er hraðbáturinn við bryggju á Höfn. Á hlið hans skráð Finnland-Ísland Viking Line.
DV-mynd Ragnar Imsl[...]

Sigldu frá Helsinki á hraðbát

Júlía Imsland, DV, Höfn:

Finnarnir Pekka og Matti Julla skipstjóri, sem sigldu frá Helsinki í Finnlandi til Íslands á 23 dögum, komu hingað til Hafnar í síðustu viku eftir giftusamlega ferð. Að vísu nokkuð slæmt veður framan af en skánaði eftir því sem þeir nálguðust Ísland.

Þeir félagar eru á 7 metra hraðbáti úr áli, Finnfaster, og komu við á 8 stöðum á leiðinni, meðal annars Kaupmannahöfn, Gautaborg og Færeyjum. Þeir hafa meðferðis skjal[...] ráðstefnu nyrstu landanna í Hels[...] í mars. Ýmsir hafa ritað nafn si[...] það og Finnarnir fara með það á f[...] hjá forseta Íslands til undirritun[...]

Destination reached on June 9, 1994. In all the hurry of writing the newspaper story in Iceland, the cre[w] of the Call of the Sagas became th[e] "Brothers Julla"!